NONVIOLENCE

Mark Kurlansky is the bestselling author of *Cod: A Biography of the Fish that Changed the World* (winner of the Glenfiddich Best Food Book Award), *The Basque History of the World, Salt: A World History, 1968: The Year that Rocked the World, The Big Oyster: A Molluscular History of New York*, a short story collection *The White Man in the Tree* and a novel *Boogaloo on 2nd Avenue* (all published by Jonathan Cape and Vintage). He lives in New York City with his wife and daughter.

MARK KURLANSKY

Nonviolence

The History of a Dangerous Idea

Foreword by His Holiness the Dalai Lama

VINTAGE BOOKS
London

Published by Vintage 2007

1 3 4 7 9 10 8 6 4 2

Copyright © Mark Kurlansky 2006

Mark Kurlansky has asserted his right under the Copyright, Designs
and Patents Act 1988 to be identified as the authors of this work

First published in the United States in 2006 by Modern Library, an
imprint of The Random House Publishing Group, a division of
Random House, Inc., New York

First published in Great Britain in 2006 by Jonathan Cape

Vintage
Random House, 20 Vauxhall Bridge Road,
London SW1V 2SA

www.vintage-books.co.uk

Addresses for companies within The Random House Group Limited
can be found at: www.randomhouse.co.uk/offices.htm

The Random House Group Limited Reg. No. 954009

A CIP catalogue record for this book
is available from the British Library

ISBN 9780099494126

The Random House Group Limited makes every effort to ensure that
the papers used in its books are made from trees that have been
legally sourced from well-managed and credibly certified forests.
Our paper procurement policy can be found at:
www.randomhouse.co.uk/paper.htm

Printed in the UK by CPI Bookmarque, Croydon, CR0 4TD

To beautiful Talia Feiga and
her entire millennial generation—
I hope you raise hell.
Nonviolently,
of course.

To kill one man is to be guilty of a capital crime, to kill ten men is to increase the guilt ten-fold, to kill a hundred men is to increase it a hundred fold. This the rulers of the earth all recognize and yet when it comes to the greatest crime— waging war on another state—they praise it!

It is clear they do not know it is wrong, for they record such deeds to be handed down to posterity; if they knew they were wrong, why should they wish to record them and have them handed down to posterity?

If a man on seeing a little black were to say it is black, but on seeing a lot of black were to say it is white, it would be clear that such a man could not distinguish black and white. Or if he were to taste a few bitter things were to pronounce them sweet, clearly he would be incapable of distinguishing between sweetness and bitterness. So those who recognize a small crime as such, but do not recognize the wickedness of the greatest crime of all—the waging of war on another state— but actually praise it—cannot distinguish right and wrong. So as to right or wrong, the rulers of the world are in confusion.

—MOZI, CHINA, CIRCA 470-391 B.C.

I find it so difficult not to hate; and when I do not hate I feel we few are so lonely in the world.

—BERTRAND RUSSELL, LETTER TO COLETTE, 1918

CONTENTS

FOREWORD

The Dalai Lama

I have worked to promote peace and nonviolence for many years because I believe that ultimately it is only through kindness and nonviolence that we human beings can create a more tranquil and happy atmosphere that will allow us to live in harmony and peace. Therefore, I am happy to see that Mark Kurlansky has wholeheartedly taken up these themes in this book.

I consider the cultivation of nonviolence and compassion as part of my daily practice. I do not think of it as something that is holy or sacred but as of practical benefit to myself. It gives me satisfaction; it gives me a sense of peace that is very helpful in maintaining sincere, genuine relationships with other people.

Mahatma Gandhi took up the ancient but powerful idea of *ahimsa*, or nonviolence, and made it familiar throughout the world. Martin Luther King Jr. followed in his footsteps. The author is correct to point out that both men were regarded with suspicion by the authorities they opposed, but ultimately both achieved far-reaching and significant changes in the societies in which they lived. I think it is important to acknowledge here that nonviolence does not mean the mere absence of violence. It is something more positive, more meaningful than that. The true expression of nonviolence is compassion, which is not just a passive emotional response, but a rational stimulus to action. To experience genuine compassion is to develop a feeling of closeness to others combined with a sense of responsibility for their welfare. This develops when we accept that other people are just like ourselves in wanting happiness and not wanting suffering.

It is my firm belief that if we adopt the right approach and make determined efforts, even in circumstances where great hostility has come about over time, trust and understanding can be restored. This is the approach I too have adopted with regard to the Chinese authorities concerning the issue of Tibet. Responding to violence with more violence is rarely appropriate. However, discussing nonviolence when things are going smoothly does not carry much weight. It is precisely when things become really difficult, urgent, and critical that we should think and act with nonviolence.

Mahatma Gandhi's great achievement was to revive and implement the ancient Indian concept of nonviolence in modern times, not only in politics, but also in day-to-day life. Another important aspect of his legacy is that he won independence for India simply by telling the truth. His practice of nonviolence depended wholly on the power of truth. The recent unprecedented fall of oppressive regimes in several parts of the world has demonstrated once more that even decades of repression cannot crush people's determination to live in freedom and dignity.

It is my hope and prayer that this book should not only attract attention, but have a profound effect on those who read it. A sign of success would be that whenever conflicts and disagreements arise, our first reaction will be to ask ourselves how we can solve them through dialogue and discussion rather than through force.

His Holiness the Dalai Lama

Nonviolence

I

IMPERFECT

BEINGS

We expect to prevail through the foolishness of preaching.

—WILLIAM LLOYD GARRISON,
Declaration of Sentiments adopted by the
Peace Convention of Boston, 1838

The first clue, lesson number one from human history on the subject of nonviolence, is that there is no word for it. The concept has been praised by every major religion. Throughout history there have been practitioners of nonviolence. Yet, while every major language has a word for violence, there is no word to express the idea of nonviolence except that it is not another idea, it is not violence. In Sanskrit, the word for violence is *himsa*, harm, and the negation of *himsa*, just as *nonviolence* is the negation of *violence*, is *ahimsa*—not doing harm. But if *ahimsa* is "not doing harm," what is it doing?

The only possible explanation for the absence of a proactive word to express nonviolence is that not only the political establishments but the cultural and intellectual establishments of all societies have viewed nonviolence as a marginal point of view, a fanciful rejection of one of society's key components, a repudiation of something important but not a serious force in itself. It is not an authentic concept but simply the abnegation of something else. It has been marginalized because it is one of the rare truly revolutionary ideas, an idea that seeks to completely change the nature of society, a threat to the established order. And it has always been treated as something profoundly dangerous.

Advocates of nonviolence—dangerous people—have been there throughout history, questioning the greatness of Caesar and Napoleon and the Founding Fathers and Roosevelt and Churchill. For every Crusade and Revolution and Civil War there have always been those who argued, with great clarity, that violence not only was immoral but that it was even a less effective means of achieving laudable goals. The case can be made that it was not the American Revolution that secured independence from Britain; it was not the Civil War that freed the slaves; and World War II did not save the Jews. But this possibility has rarely been considered, because the Caesars and Napoleons of history have always used their power to muffle the voices of those who would challenge the necessity of war—and it is these Caesars, as Napoleon observed, who get to write history. And so the ones who have killed become the ones who are revered. But there is another history that manages to survive.

It survives, but nonviolence is in fact a marginal rejection of a marginalized concept. Political theorist Hannah Arendt, in her 1969 study *On Violence,* pointed out that while it can be universally agreed that violence has been one of the primary movers of history, historians and social scientists rarely study the subject of violence. She suggested that this was because violence was such a mainstay of human activity that it was "taken for granted and therefore neglected." Violence is a fundamental of the human condition, whereas nonviolence is merely a rarified response to that reality. What does this mean? If we lived in a world that had no word for war other than *nonpeace,* what kind of world would that be? It would not necessarily be a world without war, but it would be a world that regarded war as an aberrant and insignificant activity. The widely held and seldom expressed but implicit viewpoint of most cultures is that violence is real and nonviolence is unreal. But when nonviolence becomes a reality it is a powerful force.

Nonviolence is not the same thing as pacifism, for which there are many words. Pacifism is treated almost as a psychological condition. It is a state of mind. Pacifism is passive; but nonviolence is active. Pacifism is harmless and therefore easier to accept than nonviolence, which is dangerous. When Jesus Christ said that a victim should turn the other cheek, he was preaching pacifism. But when he said that an enemy should be won over through the power of love, he was preaching nonviolence. Nonviolence, exactly like violence, is a means of persuasion, a technique for political activism, a recipe for prevailing. It requires a great deal more imagination to devise nonviolent means—boycotts, sit-ins, strikes, street theater, demonstrations—than to use force. And there is not always agreement on what constitutes violence. Some advocates of nonviolence believe that boycotts and embargoes that cause hunger and deprivation are a form of violence. Some believe that using less lethal means of force, rock throwing or rubber bullets, is a form of nonviolence. But the central belief is that forms of persuasion that do not use physical force, do not cause suffering, are more effective; and while there is often a moral argument for nonviolence, the core of the belief is political: that nonviolence is more effective than violence, that violence does not work.

Mohandas Gandhi invented a word for it, *satyagraha*, from *satya*, meaning truth. *Satyagraha*, according to Gandhi, literally means "holding on to truth" or "truth force." Interestingly, although Gandhi's teachings and techniques have had a huge impact on political activists around the world, his word for it, *satyagraha*, has never caught on.

———

All religions discuss the power of nonviolence and the evil of violence. Hinduism, which claims to be the oldest religion, though its founding date is unknown, as is its founder, does not take a clear stand on nonviolence. This ambiguity is not surprising for an ancient religion that has no central belief or official priests and has a plethora of scriptures, gods, mythologies, and cults. Hindus often repeat the aphorism *"Ahimsa paramo dharmah,"* nonviolence is the highest law, but this is not an unshakable principle of the religion. Violence is permissible in the Hindu religion, and Indra is a warlike Hindu god. But there are also many writings of Hindu wise men against violence, especially in a book known as the *Mahabharata*. Hindu sages tended to see nonviolence as an unattainable ideal. Perfect nonviolence would mean not harming any living thing. The sages encouraged vegetarianism to avoid harming animals. The Jainists, followers of a religion admired by Gandhi, keep their mouths masked to insure that they do not accidentally inhale a tiny insect. But Hinduism recognizes that even the strictest vegetarians harm plants, killing them in order to live. A saint, it is said, would live on air, but Hinduism recognizes that this is impossible. Complete *ahimsa* is not attainable. Gandhi wrote, "Nonviolence is a perfect stage. It is a goal towards which all mankind moves naturally, though unconsciously." He believed human beings were working toward perfection. Violence was a barbaric retrogressive trait that had not yet been shed. The human being who achieved complete nonviolence, according to Gandhi, would not be a saint. "He only becomes truly a man," he said.

This concept of man as an imperfect being who is obligated to strive for an unattainable perfection runs through most of human thought. The nineteenth-century French founder of the anarchist movement, Pierre-Joseph Proudhon, wrote in his 1853 *Philosophie du progrès*, "We are born perfectable, but we shall never be perfect." The often repeated argument against nonviolence, that it is in our nature to be violent—no doubt why violence deserves its own word—lacks validity in light of the ubiquitous moral argument that it is our obligation to try to be better than we are.

Hinduism and Gandhi insist that nonviolence must never come from weakness but from strength, and only the strongest and most disciplined people can hope to achieve it. Those who are incapable of defending themselves without violence, those who lack the spiritual strength to match their adversary's physical brutality, either because of their own weakness or the determined brutality of the enemy, are obligated to use physical violence for defense. In Hinduism, passive submission to brutality is usually considered a sin.

———

Whenever the Chinese denounce the pacifist tendencies in their culture, they usually blame these tendencies on Buddhism. This is because Buddhism is the only important Eastern religion in China that is of foreign origin. Buddha, the sixth-century B.C. founder, was born near the Indian-Nepalese border. If pacifism is a national weakness, many Chinese have contended, surely it is the fault of foreigners. And so Hu Shi, the Columbia University–educated Chinese scholar (1891–1962), said, "Buddhism, which dominated Chinese religious life for twenty centuries, has reinforced the peaceful tendencies of an already too peaceful people." His implication was that the rejection of violence makes people passive, and many early-twentieth-century Chinese believed their people had become too passive. This ignored the fact that most religions and philosophies that reject violence do not encourage passiveness but activism by other means—nonviolence.

Buddhism forbids the taking of life, but there seems to be a wide range of interpretations of this stance. In some countries it means vegetarianism, but in Tibet, perhaps because of a lack of vegetables, it means that animals must be slaughtered "humanely." To a Tibetan Buddhist, however, this means the opposite of what it means to a Jew. To Jews, humane slaughter is the clean slitting of the animal's throat and the removal of all blood, whereas in Tibet it means death by suffocation, to avoid the spilling of blood.

While the Buddhist interdiction on taking life was frequently interpreted in China as a condemnation of militarism, this was not the case in medieval Japan. In Japan Buddhism developed the "meditation school" commonly known as Zen. In the Middle Ages, Zen monks became warriors and monasteries became military fortresses. The original idea of Zen was the suppression of the body in order to reach a higher level of meditation. In the fourteenth century the technique was applied not only to meditation but to swordsmanship and archery. Three centuries later, Zen had become an integral part of the warrior code in Japan. This was neither the first nor last incident of a religion being perverted for military purposes.

In Buddhism, as in Hinduism, there is the notion of humans reaching higher levels, and one of the ways this is accomplished is by rendering aid to all beings. But Buddhism is not the only source of nonviolent thinking in China. The position on war and nonviolence in Confucianism, a belief system developed in China from 722 to 484 B.C., is even more vague than in Hinduism. It is not even clear that Confucianism is a religion. Many prefer to describe it as a moral philosophy. Nor is there agreement on the extent of the role of Confucius, whose real name was Kong Fuzi, a contemporary of Buddha, who lived between 551 and 479 B.C. *The Analects*, a compilation of Confucius's sayings that was assembled long after his death, defined the function of government as providing food and troops and earning the people's confidence. Asked which could be suspended in hard times, he answered, "Dispense with the troops." This idea that military is essential to government but less essential than other functions runs throughout *The Analects*.

Confucius was not a pacifist, nor did he teach the power of non-violence. But *The Analects* also at times rejects the notion of state violence, saying, "If good men were to administer the government for a hundred years, violence could be overcome and capital punishment dispensed with." And when the question comes up of how to deal with neighboring barbarians, the standard rationale for military campaigns in China, the reply in *The Analects* is: "If the distant peoples do not submit, then build up culture and character and so win them, and when they have been won give them security." It is a succinct statement of the nonviolent approach to political activism.

But the strongest Chinese stand on nonviolence came in opposition to Confucius, from a man named Mozi, who lived from about 470 to 390 B.C. Mozi frequently attacked Confucianists for being aristocrats, which has led some scholars to conclude that he came from a class of slaves. But like other rebels, including Jesus and Gandhi, he may have chosen to throw in with the poorest class as a protest against their unfair treatment. While Confucius was a voice of the establishment, Mozi was a rebel. While Confucius envisioned a hierarchy of love in which the greatest affection was given to family, Mozi called for universal love, *chien ai*, and emphasized helping the poor. Mozi described the concept of *chien ai*: "He throws me a peach, I return him a plum."

Mozi saw this concept of mutual love, *chien ai*, as the key to righting the world's ills.

Whence come disorders? They arise from lack of mutual love. The son loves himself and does not love his father and so cheats his father for his own gain; the younger brother loves himself and does not love his father and so cheats his elder brother for his own gain. The same applies to the state officers and their overlords. This is what the world calls disorder. In the same way the father loves himself and not his son and cheats his son for his own profit, and so likewise with the elder brother and the overlord. This all comes from the lack of mutual love. Their case is the same as that of robbers and brigands who likewise love their own households, but not the homes of others and so rob others' homes for the benefit of their own. Like unto these, too, are state officers and princes who make war on other countries—because they love their own country but not other countries, and so seek to profit their own country at the expense of others. The ultimate cause of all disorders in the world is the lack of mutual love.

Mozi goes on to make a point that was later voiced in Judaism by the first-century A.D. rabbi Hillel and reiterated by his contemporary, Jesus, who called it the Golden Rule. Mozi wrote:

For if every man were to regard the persons of others as his own person, who would inflict pain and injury on others? If they regarded the homes of others as their own homes, who would rob others' homes? Thus in that case there would be no brigands or robbers. If the princes regarded other countries as their own, who would wage war on other countries? Thus in that case there would be no more war.

Chinese comes closer than most languages to a word for nonviolence. In Taoism there is a concept embodied in the word *teh*. Not exactly nonviolence, which is an active force, *teh* is the virtue of not fighting—nonviolence is the path to *teh*.

Taoism is centered on the fifth-century B.C. teachings of Lao Tsu, who is thought to be the author of the *Tao te ching*, The Cannon of the Way and Virtue. *Tao* itself is an untranslatable word, often mentioned in *The Analects*. It is a balancing force sometimes said to be what keeps nature from tumbling into chaos. It says in the *Tao te ching*, "The ruler imbued with the Tao will not use the force of arms to subdue other countries." But it adds that a country should have a military force for defense and that its preparedness will be a deterrent. The military should be "ready but not boastful." This half-road to nonviolence is not nonviolence at all, since all of history shows that nations who build military forces as deterrents eventually use them—a disturbing lesson in an age of "nuclear deterrents."

But there is in Tao, as in Hinduism, the notion that human beings evolve and the more highly evolved human beings do not need physical violence. "The skillful knight is not warlike. The skilled strategist is never angry. He who is skilled in overcoming his enemies does not join battle."

In Taoism *teh* is a perfection of nature, and, as in Hinduism, it is something few people have the strength and character to live up to. The concept is echoed in Christianity by such notions as the meek being blessed and the last being first. *Teh* holds that:

> In nature the softest overcomes the strongest. There is nothing in the world so weak as water. But nothing can surpass it in attacking the hard and strong; there is no way to alter it. Hence weakness overcomes strength, softness overcomes hardness. The world knows this but is unable to practice it.

Eastern religions, which Westerners tend to regard as ethereal and only workable for the dreamiest of idealists, actually have a pragmatic side. They recognize that violence is wrong, that nonviolence is the path that ought to be taken, but they also recognize that humans are weak and imperfect and that only a few of the most evolved and extraordinary among us will choose that path and stay with it.

———

Judaism, a religion that is more than 5,700 years old, has many layers of both laws and commentaries on those laws. It is full of seeming contradictions, including on the subject of violence. Rabbis attempt to resolve the contradictions by ascribing priorities— certain writings are more important than others, some doctrines, some practices, some beliefs take precedence over others. Of course the arguments about which writings take precedence are without end. In Judaism there is usually room for arguments, but there are a few inviolable laws. Monotheism is the central tenet of the religion and there are no exceptions or variations, nor is any form of idolatry tolerated. It is also universally accepted that the ten commandments that are said to have been handed to Moses by God on Mount Sinai are a central and leading set of nonnegotiable laws. The first of these commandments is monotheism and the second forbids idolatry. The sixth commandment is "You shall not kill." It is one of the shortest commandments and offers no commentary, explanation, or variations. It does not say, as many Jews claim, "except in self-defense," nor does it say "except when absolutely necessary." It is one of the most plain declarative sentences in the Bible. But those who wish to kill can take refuge in lesser writings. The Old Testament is full of accounts of warfare and even justifications for them. This does not change the fact that the central law states "no killing." Throughout the rest of the Bible, among all the battles and bloodshed, are other messages. The dictum in the book of Leviticus, "Love your neighbor as yourself," is also considered central to the religion.

The ancient Jews did engage in warfare, but they apparently never felt comfortable about it. Unlike so many modern cultures, they did not celebrate military victories. The only holiday on the Jewish calendar that celebrates a military triumph is Chanukah. It is a postbiblical holiday celebrating the 166 B.C. victory of a guerrilla army led by the Maccabees against the Seleucid rulers of Palestine who, with the support of some Jews, had tried to dilute traditional Jewish practice. Rabbis were never comfortable with this holiday, and the writings that record it were not kept with sacred text and have only survived in Greek translation, the language of the defeated. Chanukah was always a minor holiday of very limited religious significance until modern times, when two things happened to change its role. In the 1890s, with the growth of Zionism, Chanukah was promoted because it celebrated the Jewish military conquest of Jerusalem. Like the Zen monks, the Zionists knew how to use religion in the quest for political power. Today in Israel it is virtually a political holiday.

Chanukah's popularity continued to grow, though it is still not considered a religious holiday, and it has been given new importance in modern times by retail merchants eager to sell gift items to Jews during the Christmas season. The traditional time of year for giving children gifts in the Jewish calendar used to be Purim, which falls at the end of winter. While not celebrating a military victory, this holiday is also bloodied by the hanging of the wicked Haman and his cohorts at the city gates and the slaughter of 75,000 Persians. Centuries of commentaries have discussed the unseemly grisliness of this story. But while most Jewish holidays are somber, Purim is intended to be a time for silliness, a bit like the pre-Lenten Mardi Gras in Catholicism. Drunkenness is encouraged, as is ridiculing revered scholars. The story of the book of Esther is retold on Purim intentionally as a farcical, overblown melodrama in which the good guys are cheered and the bad guys booed. Scholars and rabbis point out that "God is not present" in the story of Purim. The book of Esther is the only book in the Old Testament, aside from the love poem Song of Songs, in which God never appears. The characters do not pray, they do not ask God's help. God is not involved in this bloody operation. It has already been made plain that God does not want people to kill each other.

Generally Jewish holidays reject such violence. On Yom Kippur violence is among the sins for which to atone. On Passover, which celebrates Moses leading the Hebrews out of slavery in Egypt, there is a moment of sorrow for the Egyptians, the enemy who drowned trying to pursue the Hebrews across the Red Sea. Jews are instructed every year not to hate the Egyptians. It is a fundamental tenet of Judaism that you should not hate your enemies.

Judaism, too, teaches of the possibility of perfection. Someday, it is said, the perfect human, the Messiah, will come and show mankind the way to perfection. By tradition, the Jews were to return to Israel only when the Messiah appeared, not following World War II. Reform Judaism does not predict a Messiah but an entire messianic age. According to the angry prophet Isaiah, at some point in the future, when God is finally listened to, nations "shall beat their swords into ploughshares and their spears into pruning hooks, nation shall not lift sword against nation, nor ever again be trained for war."

———

Though most religions shun warfare and hold nonviolence as the only moral route toward political change, religion and its language have been co-opted by the violent people who have been governing societies. If someone were to come along who would not compromise, a rebel who insisted on taking the only moral path, rejecting violence in all its forms, such a person would seem so menacing that he would be killed, and after his death he would be canonized or deified, because a saint is less dangerous than a rebel. This has happened numerous times, but the first prominent example was a Jew named Jesus.

II

THE PROBLEM WITH STATES

If the force of arms is considered the only means of authority,
it is not an auspicious instrument.

—LAO TSU,
the *Tao te ching*,
fifth century B.C.

Jesus, like Mohammed after him, looked at the great complexity of Jewish law that had been layered over millennia and said that implanted in the law were certain clear precepts of right and wrong. Others offered the same clarity. Hillel, a Babylonian Jew who lived at about the same time as Jesus, also preached a message of simplicity. Like Mozi, Hillel was said to have come from particularly humble origins. He studied by climbing to the roof of the school, literally eavesdropping on lessons, because he had no money to register as a student. Hillel became the head rabbi of Palestine, from which position he constantly wrestled with the conservative rabbinate. In a stance that is unusual even today, he was extremely open to converts. One aspiring convert, apparently frustrated with the verbosity of Jewish law, challenged him to recite the Torah while standing on one foot. Hillel responded, "What is hateful to you, do not do to your fellow. The rest is commentary on this. Go and study." Hillel's followers became the dominant political force among the Jews.

Jesus and his followers were clearly influenced by Hillel. Hillel's summary of the Torah became Jesus' "Golden Rule" in the Sermon on the Mount: "In everything, treat people the same way you want them to treat you, for this is the Law and the Prophets." Jesus taught the doctrines of Judaism. Where he differed was in priorities. As with traditional Judaism, his first priority was the love of God. But his number two was the love of man. Jesus believed that love should be given to all fellow humans unconditionally. In the Sermon on the Mount, Jesus' most succinct sermon, which he delivered seated, in the traditional manner of a rabbi, he made clear that he did not want to reject Judaism but to revive it and have its most important laws more rigorously observed. "Do not think that I came to abolish the Law or the Prophets; I did not come to abolish but to fulfill." His first example was the interdiction against killing. But he went even further. Even being angry at a fellow human being was a sin. In Jesus' view of Jewish law, there was no room for violence of any kind, even emotional violence, and weapons, military, and war were clearly illegal. The righteous person who walked in God's path loved everyone, even his enemies.

Jesus was seen as dangerous because he rejected not only warfare and killing but any kind of force. Those in authority saw this as a challenge. How could there be authority without force? This was trouble for the rabbinate, and was even more trouble for the military occupiers, the Romans. Jesus built a following that was attracted to his uncompromising point of view—the kind of people who are called troublemakers. He was tortured to death by the Romans in a manner so grisly and violent, it was surely designed to repel his followers. But they insisted that Jesus had died forgiving his torturers.

Death by crucifixion is believed to be a Phoenician invention. Unarguably a horrifying death, it was thought by the Romans to be a humiliating and degrading one as well, and they did not use it on Roman citizens. Those first Christians would not have used the symbol of a cross, a weapon of violence, much less a crucifix, which was a depiction of violent death. They were led by a fisherman, and their symbol was a fish.

The early Christians persisted in an uncompromising and narrow interpretation of Jewish law. In the book of Matthew is written, "You have heard it said: An eye for an eye, and a tooth for a tooth. But I say unto you, resist not him that is evil." The rejected eye-for-an-eye formula is not a peripheral piece of Jewish commentary, it is from the book of Exodus. Major Old Testament figures, including not only David but Samson, Joshua, and Gideon, were military men. But slowly the idea emerged among the followers of Jesus that they should hold Jewish law to a higher standard and that though warfare had been tolerated, it would be no longer.

A split, the first and probably the most important of many schisms in Christianity, occurred between Jesus' disciples Peter and Paul. Paul, whose original name was Saul, and Peter, who was originally named Simon, were both Jewish. But Paul, unlike Peter, was not one of Jesus' entourage and never knew him. While Peter was a fisherman in Galilee, Paul was a religious scholar from Asia Minor. And yet it was Peter, the fisherman, who wanted the followers of Jesus to remain Jewish and apply Jesus' teachings to the perfection of Judaism. Paul, the Hebraic scholar, wanted to open up Christianity to the world, pursuing converts wherever they were found, a most un-Jewish approach.

Under Paul's influence the Christians moved further away from the body of Judaism, further away from everyone. They became an odd and distinct cult on the outer margins of society, uncompromisingly dedicated to pacifism. Theirs was a unique antiwar posture. Even the pious and spartan Jewish sect known as the Essenes did not entirely denounce weapons.

The early Christians are the earliest known group that renounced warfare in all its forms and rejected all its institutions. This small and original group was devoted to antimilitarism, another concept, like nonviolence, that has no positive word. This antimilitarism was never expressed by Jesus, who, in fact, did not much address the issue of warfare, though he did denounce the violent overthrow of the Romans. Warmongering Christian fundamentalists have always clung to the absence of a specific stand on warfare, ignoring the obvious, which is that the wholesale institutionalized slaughter of fellow human beings is clearly a violation of the precise and literal teachings of Jesus. In the days of the great Western debate on slavery, slave owners used a similar argument—that Jesus had not said anything about slavery. But obviously the buying and selling of human beings would not constitute treating others as you would have them treat you.

For 284 years, roughly the same span of time as from the end of Louis XIV's reign in France to the beginning of the twenty-first century, Christians remained an antiwar cult. Christian writers emphasized the incompatibility of warfare with Christian teachings. Some characterized warfare as the work of evil spirits and weapons as cursed. They labeled the taking of human life in warfare murder. The Jewish War of A.D. 66–71 was viewed as God's punishment of the Jews for sinful ways, and the pursuit of war, and by extension the pursuit of power politics, was said to be activity for "the Gentiles," unworthy of a Christian. They attacked the pomp of Rome as a glorification of warfare.

The first-century Christian writer Ignatius called for an abolition of warfare. This would happen, according to him and other Christian writers, once the world embraced the teachings of Jesus Christ—to love one's enemies, to do good even to those who do evil, to respond to evil with goodness. Such determined love and goodness was not meant to be pacifistic but a program for actively fighting evil. Given their stance against soldiers and soldiering, even against police work, it is striking that Christians sought and got converts among the Roman Legions. Some historians believe that it was converted Roman soldiers who first brought Christianity to Britain.

Origenes Adamantius, popularly known as Origen, the second-to-third-century Christian philosopher from Alexandria, clearly stated, "We Christians do not become fellow soldiers with the Emperor, even if he presses for this." Christians would be loyal to the emperor, but they would not fight his wars. According to Origen, a Christian might pray for the success of a military state, even pray for the success of a military campaign, but could never participate in the military or in the government of a state that used military power. He did not condemn the military but only believed that it was forbidden for a Christian to participate. Christianity was about the promotion of love, and early Christians believed that love and killing were incompatible.

Though no one doubted Origen's sincerity—after all, he had castrated himself in pursuit of personal purity—his was a dangerous position in a militarized state. Like many subsequent states, the Roman Empire was so invested in its military might that it found it difficult to conceive of a loyal citizen who would not participate in the central program—warfare. Origen understood this, since his father had been put to death for beliefs similar to his own. Origen himself, the most influential Christian thinker of his time, author of some 800 works, was imprisoned and tortured and died from his mistreatment shortly after being released, in about A.D. 254.

Not all Christians were good Christians. Some were described by other Christians as "behaving like Gentiles." Starting in the mid-second century, under the reign of Marcus Aurelius, a notable persecutor of Christians, some Christians did become soldiers and others became magistrates. Apparently there was an attempt to force Christians into the military, for about this time the first evidence is found of Christians refusing to serve—Western history's first conscientious objectors.

In general, Christians were becoming more troublesome rather than less. Their habit of seeking converts among the legions was a direct threat to empire building. Most soldiers, upon converting, refused to continue military service. Tertullian, a Roman centurian's son who converted to Christianity in 197, openly spoke of converting soldiers so that they would refuse to fight.

Active practitioners of nonviolence are always seen as a threat, a direct menace, to the state. The state maintains the right to kill as its exclusive and jealously guarded privilege. Nothing makes this more clear than capital punishment, which argues that killing is wrong and so the state must kill killers. Mozi understood that the state's desire to kill had to do with power. He wrote: "Like unto these, too, are state officers and princes who make war on other countries—because they love their own country but not other countries, and so seek to profit their own country at the expense of others."

To those who govern, the citizen who questions the right of the state to kill is attempting to impinge on the government's ability to further the nation's interests at the expense of other, hopefully weaker, states. Thus the nonviolent activist is seen as a threat to the state.

In 274 in Numidia, where Algeria is today, a soldier of the Roman Empire, Fabius Victor, had a son named Maximilianus who, like all sons of military men, was drafted into service when he turned twenty-one. But instead of reporting for duty, he told Cassius Dion, the proconsul of Africa, that he was a Christian and therefore could not enter the military because he owed his first duty to the teachings of Christ. His father the veteran did not entirely agree with his son but was supportive. Maximilianus insisted, "I cannot serve as a soldier. I cannot do evil. I am a Christian." Dion countered that there were Christians in military service throughout the empire. "What evil do they do who serve?"

To which Maximilianus answered, "You know what they do."

The young man was taken away and executed.

Historians have had some problems with the story, including a lack of records to verify the existence of a proconsul in Numidia named Cassius Dion, but the stance and punishment of Maximilianus is recorded and he is remembered as the first martyred conscientious objector.

Toward the end of the third century, the Roman military further distanced itself from Christianity by requiring officers to practice the Roman religion, which to Christians was paganism. This led to more Christian officers resigning from the military. The desertion of Christians and Christian converts was a growing problem.

Then came the triumph of Christianity, a calamity from which the Church has never recovered.

———

Constantine I, son of Constantinus Chlorus, caesar of the Roman Empire, the number-two position of power beneath emperor, struggled to advance to emperor. In 312 Constantine's army was to fight a decisive battle against his principle competitor, Maxentius, at the Milvian Bridge over the Tiber. The early fourth century was a time of great belief in magic, spells, dark powers, and unseen forces, and Constantine, according to the chronicler Eusebius, had felt the need for something greater than military might to defeat the rulers of Rome, whom he was convinced had rallied forces of black magic. Constantine had a dream in which Christ had appeared, commanding him to carry the sign of the cross into battle. By this time Constantine had numerous Christian soldiers in his ranks, and for the first time in history they went into battle with an emblem of Christianity, the cross, painted on their shields. Just a generation earlier, to have placed a symbol of Christianity on a weapon would have been an outrage for Romans and an unthinkable blasphemy for Christians. Before the battle, Constantine was said to have seen a flaming cross in the sky with the words "In this sign thou shalt conquer," words that were in complete contradiction to Christianity and would have been unutterable for Jesus.

Unfortunately for Christianity, Constantine and his Christian warriors won the battle, establishing him as ruler of the western half of the Roman Empire, but also establishing a new role for the Christian and for Christ, a God who now would not only sanction killing but would take sides to help one band of killers triumph over another. The following year, Constantine met in Milan with his co-emperor in the East, Licinius, to issue the Edict of Milan. The Edict of Milan reiterated what Galerius, yet another rival to the throne, had already decreed in 309, the year before he died, that Christianity would no longer be illegal in the Roman Empire. It did not, as is commonly believed, make Christianity the official state religion.

The move was cheered by Christians, even those who had reservations about the new Christian soldier, because the edict meant an end to their persecution. In the remaining twenty-five years of his life, Constantine built grand churches, including three of the grandest—St. Peter's in Rome, Hagia Sophia in Constantinople, and the Holy Sepulchre in Jerusalem. Constantine became the Christian emperor, the defender of Christianity, and, as such, forever changed the character of the religion as he promoted it and used it to solidify his power. Whether he himself embraced the religion or simply used it politically has been debated by historians ever since. He empowered the Church as an instrument of statecraft, spending a large part of state funds in the establishment and control of clergy. He declared his prayer day, Sunday, as an official day of rest and prayer for the entire empire. By enforcing this one edict, the Church became a major force in everyone's daily lives.

One of history's greatest lessons is that once the state embraces a religion, the nature of that religion changes radically. It loses its nonviolent component and becomes a force for war rather than peace. The state must make war, because without war it would have to drop its power politics and renege on its mission to seek advantage over other nations, enhancing itself at the expense of others. And so a religion that is in the service of a state is a religion that not only accepts war but prays for victory. From Constantine to the Crusaders to the contemporary American Christian right, people who call themselves Christians have betrayed the teachings of Jesus while using His name in the pursuit of political power. But this is not an exclusively Christian phenomenon. Hinduism, Buddhism, Islam, Judaism—all the great religions have been betrayed in the hands of people seeking political power and have been defiled and disgraced in the hands of nation-states.

Christianity began to change immediately upon the warm embrace of Constantine's empire. The year after Christianity became a legal religion, a Church synod at Arles ruled that conscientious objection would only be acceptable in time of war. Any Christian who refused military service in peacetime was to be excommunicated from the Church. Significantly, during this period, in A.D. 326, the empress Helena announced that buried deep in the hillside at Calvary she had discovered the "True Cross," the gruesome implement of torture used to kill Jesus, supposedly made of wood from the Garden of Eden. And this bloodstained implement of violence, the image of which had been woven into banners and painted on shields at the Milvian Bridge, now became the official symbol of the Church.

A permanent split developed between the impassioned followers of Jesus and the official Church, which, having a pro-state bias, compromised on principles with legalistic arguments to allow states to continue functioning the way they always had. The assumption was that this was the only way a state could operate.

Some Christians continued to refuse military service. In 336 another son of a soldier suddenly put down his arms before a battle and refused to fight. The young man, Martin, had served in the military for two years after his conversion to Christianity. One day Martin said, "I am a soldier of Christ. I cannot fight." He was accused of cowardice, to which he responded by offering to go unarmed in front of the troops onto the battlefield. The emperor decided a fitting end to Martin would be to take him up on his offer, but before this could happen peace was negotiated with the Gauls.

The battle never took place, leaving Martin to die a natural death sixty-one years later at the age of eighty-one.

But others refused service, too, including Martin's friend Victricius. The Church addressed this Christian urge toward conscientious objection later in the century, declaring that a Christian who had shed blood was not eligible for communion for three years. Thus did the Church acknowledge an objection to warfare, but not an insurmountable one. Then in the fifth century an Algerian bishop, Augustine of Hippo, wrote the enduring apologia for murder on the battlefield, the concept of "just war." Augustine, considered one of the fathers of the Catholic Church, declared that the validity of war was a question of inner motive. If a pious man believed in a just cause and truly loved his enemies, it was permissible to go to war and to kill the enemies he loved because he was doing it in a high minded way.

———

The Catholic Church turned Maximilianus into Saint Maximilian of Tebessa. Jesus Christ, whose teachings had been dismantled by everyone from Constantine to Augustine, was much more than a saint; Christians declared him a deity, the son of God. Martin, who refused to go into battle against the Gauls, is now Saint Martin of Tours. Martin did not really qualify for sainthood, since, according to the original rules of the Catholic Church, one of the requirements was martyrdom. Martin would have been a fine saint if it weren't for the last-minute peace with the Gauls. He would have marched unarmed across the field, been cut down and chopped up for sainthood. The later Church, not the one Martin knew, needed martyrs, because extolling martyrdom is a way of promoting warfare—the glory of being slaughtered. Needing Martin safely as a saint on their side and not as an unclaimed rebel conscientious objector, the Church turned Martin of Tours into the first unmartyred Catholic saint.

Saint Martin has become a kind of military figure, usually portrayed in armor. The U.S. Army Quartermasters Corps awards a medal named after him, "The military order of Saint Martin." Saint

Martin is supposed to have died on November 11, 397. Historians say that the day is uncertain, but the date has taken on absolute certainty as the Feast of Saint Martin, because it coincides with the date of the armistice ending World War I. It is difficult to know what to do with rebels, but saints have a thousand uses.

III

THE KILLER
PEACE MOVEMENT

[We call for peace] in the name of God, since without peace
no one will see God.

—Peace meeting at Le Puy, 994

The ideology of warfare that has been repeatedly invoked for the past thousand years of Western history grew out of Augustine's thesis of just war in the fifth century and continued to be developed to its complete expression in Pope Urban II's propaganda campaign launching the first Crusade at the end of the eleventh century.

Simply stated in the terms of the American western, one of the great cultural institutions for fostering violence, the world is made up of good guys and bad guys, and the good guys have to shoot the bad guys for everyone's well-being. Once this was established, the state had only to declare its proposed victim a bad guy to justify a war.

If Christianity was initially polluted by the state, in the second phase the state was polluted by Christianity. Once the religion began working with the state and became involved in the state's business, it was involved in warfare. Augustine provided the theology to explain this unexplainable contradiction. But in the process, the role of the Christian Church was changing. From a moral guide on the periphery of events, it moved to the epicenter of power politics.

The state jealously guards the right to make war because this prerogative is a source of power. Once Christianity became interested in power, the Church became competitive with states. If kings derived their power from the right to declare war, the clergy would challenge that power with the right to declare peace. And so began a power struggle in which a peace movement known as *pax dei*, the Peace of God, led the world into the ruthless and violent wars known as the Crusades.

The Church engaged in this power struggle for some time before the late tenth century, when the Peace of God came into being as a recognized movement. Officially it seems to have begun at a 975 meeting in an open field outside the city walls of Le Puy, which is today in France. From the beginning the movement was not really about peace. The meeting was called by the Church to discuss the raiding and looting of Church holdings by noblemen leading peasant armies. The noblemen were forced to take an oath that they would no longer commit such acts of aggression against Church property. If they broke their oath the penalty was excommunication. But the threat was also backed up by considerable military might.

The meeting in Le Puy was considered a great success, and others followed on the same model. To attack Church property—buildings, clergy, livestock, crops, olive trees, peasants while harvesting—was a crime against the *pax dei*. Sometimes widows, orphans, and others considered to be defenseless were also included in the protection of the Peace of God. Some fifty years later, either coming out of the Peace of God movement or running alongside it—historians disagree on this point—a movement arose called *truega dei*, the Truce of God.

A truce is not a peace. The Truce of God movement did little to end war but did a great deal to establish the power of the Church. The Church declared a moratorium on warfare during holy days just as it had ordered abstinence from sex and red meat on those days. Since holy days made up more than half the days of the year, including every Sunday, the Truce of God meant that by Church orders, under threat of excommunication, a king who was engaged in a war had to constantly lay down his sword for a day or two in mid-campaign. This alone made the Church far more powerful than it had ever been.

This strengthened Church enforced its authority to stop violence not only by the threat of excommunication but by mobilizing powerful armies, which it used to wage war to chastise "peacebreakers" who had violated Church truce days. Among the combatants in these Church armies were clergymen killing for peace, a just war. These armies, sometimes called peace militias, unleashed terror on populations, razing whole castles and slaughtering peasants who had fled to the protection of the castle ramparts. In one incident the peace militia massacred fourteen hundred men and women. Sometimes peace-breaking lords would retaliate and slaughter hundreds of clergy. They were two opposing dominions, the religious and the secular, rallying military might to fight for power.

How far Christianity had come from the time when a Christian, by definition, took no part in warfare. Until the eighth century, clerics had been barred from combat, even in a "just war." After that they were allowed to accompany troops to celebrate mass, hear confessions, and perform other priestly functions. Even in the eleventh century clerics were forbidden to bear arms. But as the Church asserted its power, it took on more military functions, provisioning armies, conscripting soldiers, and finally leading campaigns. Some priests went into battle with clubs, because they believed it was unchristian to wield a sword. After all, Augustine had argued that Jesus was not really talking about loving one's enemies but simply loving the reflection of God that was within them. A priest could love the reflection of God within someone and still club that person to death, which was more moral than stabbing or chopping him to death.

By the early eleventh century, Christians were venturing the opinion that not only was violence acceptable but that killing pleased God when done in the cause of the Church. The killing of "false Christians" pleased God and was not to be considered homicide. Then in 1063 Pope Alexander II wrote Wilfred, Archbishop of Narbonne, that there were two exceptional circumstances under which killing people was permissible: in punishing crimes and in stopping aggression. He offered an example of stopping aggression: stopping the Saracens, a somewhat pejorative Western term for Muslims. In fact, by the eleventh century a term was invented for killing non-Christians, since this was not to be considered homicide. It was *malicide,* Latin for the killing of a bad person.

———

Islam—the root of the word is *salam,* peace—was founded by the seventh-century A.D. prophet Mohammed and takes much from Judaism, including a few of the many dietary laws and numerous historical characters, such as Abraham, Moses, and David. Islam also recognizes the Christian founder, Jesus, but does not accept his divinity. The prophet Mohammed was troubled by the growing materialism of his people. Like Jesus, he had no intention of founding a new religion but wanted to bring the spiritual values of monotheism to Arabs. Mohammed, who was not literate, claimed periodically to have revelations, each one about a paragraph in length, and after twenty-one years these revelations were put together in a book, considered a masterpiece of Arab writing, called the Quran. Central to the Quran is the building of communities with a just distribution of wealth. Mohammed's approach shunned abstract debate and encouraged pragmatic solutions. He always emphasized negotiating solutions, and by tradition there is tremendous emphasis on negotiation in Muslim history. Mohammed's attempt at a perfect society in Mecca enforced a complete ban on violence, which made Mecca prosper as a center of trade. During the *hajj*, the required pilgrimage to Mecca, the faithful Muslim was not allowed to carry weapons, even for hunting, nor to commit any violence, including words spoken in anger.

Islam, an unusually open faith whose early adherents came from many backgrounds, including Judaism, began to change after 622, when Mohammed and his followers moved from Mecca to Yathrib, a town 250 miles to the north, which was renamed *al-Medinah*—the city. One of the most bitter disappointments of Mohammed's life was that the Jews of Medina, apparently more tradition-bound than those he had known before, refused to accept him as a prophet. At this point Jewish prophecy was already several thousand years old and they viewed a man who claimed to get messages from God with the same suspicion that most people today would. Mohammed physically turned the prayer meeting around so that worshipers now faced Mecca rather than Jerusalem.

But the establishment of Medina had an effect on Islam not unlike that of Constantine and Rome on Christianity. It was not that Mohammed was interested in conquest and empire like Constantine, but Medina had become, in effect, a state—territory that had to be defended when it was attacked by men from Mecca who vowed to destroy it, and in 625 they almost succeeded. The defense of Medina, in several major battles, began Islamic military history and included the first Muslim-Jewish conflict, in which Mohammed massacred an armed Jewish group that rose against him. By the seventh century it was already an old pattern: the religious doctrine of peace meets the power politics of state, the rules are bent for the "just war," and once the first few doses are administered the state becomes an addict that will tell any lie to get its narcotic. War is simply the means. The real narcotic is power. As Hungarian writer György Konrád said of the United States and the Soviet Union in the 1980s, "Men can invent few libidinous fantasies more enjoyable than those of world domination." The African-American poet Langston Hughes called the leading nations "the nymphomaniacs of power."

Mohammed, so the rationale goes, had to defend Medina. The Quran says, "Permission to take up arms is hereby given to those who have been attacked because they have been wronged." There is no provision, however, for preemptive strikes. Wars of aggression are immoral and forbidden. Wars to spread the teachings of Islam are also not permitted. But seventh century Islamic warfare was never justified on the grounds of building an empire or even spreading Islam. Mohammed did not want to convert Jews and Christians, since they were already believers in the one God. Islam teaches respect for the revelations of other prophets in other groups.

But there is always a way to argue that a war is a case of self-defense. The Jewish rebels had to be slaughtered to teach a lesson and stop the uprisings. Alliances needed to be formed for defense, and then obligations had to be met. Along the way it was noticed that warfare gave a common purpose and united Arab peoples, though sometimes an individual warlord simply coveted a strip of land.

Mohammed himself fought nineteen campaigns, but he taught that war was only a last resort and that God blessed those who took a nonviolent rather than a violent path—"God grants to gentleness what he does not grant to violence."

After the death of the prophet Mohammed in 632, Islam's character changed even more rapidly than that of Christianity following Jesus' death. Mohammed's Muslims took Palestine in 636 and conquered Jerusalem in 638, and then went on to Syria, Egypt, and North Africa. None of which erases one of the most important passages of the Quran: "Whoever kills a human being should be looked upon as though he had killed all mankind."

War between Muslims, like war between Christians, was both commonplace and morally unacceptable, except, again, in cases of defense. The Quran states: "Do not yield to the unbelievers. Fight them strenuously." For centuries this quote has been used to imply that Muslims had an obligation to go to war with non-Muslims. But many Islamic scholars have stated that the Quran seems to have meant by the word "fight" the use of intellectual persuasion, since the suggested tool is the Quran and it can be assumed that a good Muslim is not being told to whack the unbeliever over the head with a large copy of the sacred book.

In Islam, too, there is the concept of striving toward perfection. The Quran says that a good Muslim must "strive for the cause of Allah." In Arabic this striving or struggle is called *jihad. Jihad* originally meant striving with great intensity. But this striving was meant to be an internal struggle to become the perfect Muslim that God—Allah—wanted each Muslim to be. Some scholars have even argued that when the Quran speaks of conquering unbelievers with *jihad,* it is saying to persuade them with the force of argument, and thus *jihad* means nonviolent activism. This is why numerous notable Islamic clerics have said that the prophet Jesus also instructed his followers to wage *jihad.*

Islamic scholars have always debated the meaning of the thirty-five references to *jihad* in the Quran. But as medieval Muslims became engaged in a series of difficult wars, the word *jihad* began to be used to denote the struggle to prevail militarily in place of the original word for such a physical battle, *qital.* All successful leaders understand the importance of words, and it seemed a good Muslim would fight harder if the struggle were called *jihad* rather than *qital.* After the death of Mohammed, Muslims began speaking of two kinds of *jihad—al jihad al akbar,* greater *jihad,* and *al jihad al asghar,* lesser *jihad.* Greater *jihad* was the struggle to be a pure and good person, while lesser *jihad* referred to armed struggle.

A century after the death of Mohammed, Muslim armies had crossed the Pyrenees and penetrated deeply into Europe, controlling much of the Mediterranean and ranging east as far as India. But any un-Mohammedan dreams of Islamic world dominance they may have had faded early in the eighth century when it became apparent that they were not going to be able to conquer Constantinople. After that, Islamic legal writing increasingly accepted the notion of the infidel state. Islam developed a peace concept that brought no more peace than had the Christian Peace of God. It was called *dar al-suhl*, the House of Peace, and it meant that Muslim states could honorably live in peace with non-Muslim neighbors.

The House of Peace easily crumbled. When the Byzantine Empire began aggressively moving east in the tenth century, Sayf al-Dawla, a celebrated Shi'ite ruler of the northern Iraqi Hamdanid dynasty, declared an annual *jihad* against Byzantium. Later in the same century, the Hamdanid rulers in what is now Turkey began delivering celebrated *jihad* sermons in ornate and rhymed phrases. The Hamdanid court had been famous for its poets. Now, using such poetic devices as alliteration and repetition, these sermons on *jihad*, considered a high point in the tradition of Arab oratory, challenged Muslims to holy war. The sermons were said to be so beautiful that people were moved to tears, and also to kill. They became an enduring model for Islamic war propaganda.

But by the late eleventh century, when the Arabs were about to face their greatest military challenge, Arab writers were complaining that nobody seemed interested in *jihad* anymore.

———

By this time in Europe, the Church had so skillfully perfected the ability to make peace that it held the power to make war as well. Pope Gregory VII took the next step, demanding that secular princes do their part to furnish the Church with fighters and weapons whenever he needed them "in the service of Saint Peter."

The eleventh century was an impassioned time. The excitement about the new millennium in 2000 focused largely on whether computers would continue working; but in the year 1000, Europe was focused on the first millennium of Christ. All events—storms, comets, wars, floods, epidemics—were given momentous significance. God's wrath was seen at every turn. Christians seemed to believe they had strayed very far from God's will. People flocked to the churches and clung to the words of the clergy. But the Church was not talking about restoring the teachings of Jesus Christ; rather, the clergy spoke of defending the Church against its enemies. In the course of the first century of the new millennium, the Church, the most powerful institution in Europe, had overwhelmed European princes, to the point where its only remaining challengers were the Saracens. The fact that these Saracens were in decline made battling them all the more irresistible. It was time for a Christian lesser *jihad.*

Two popes later, Urban II, a Frenchman, launched the Crusades in a rousing speech at a "peace council" at Clermont in 1095. Urban declared that the pope could now define both the peace and truce of God for all of Christendom. Christendom itself, the idea that all Christians were a cohesive singular force, was a relatively new concept that the papacy had been nurturing for the past two centuries. He then declared peace in the West, freeing Christians to launch a "holy war" to liberate Jerusalem from the Saracens—granting Peace of God protection to the families and property of those who marched off to the Holy War. Now the doctrine of the peace movement—that without peace, God could not be served—was slightly bent to a new truth: that without peace in the West, the Church could not successfully launch its expedition to the East. This expedition was to be one final war in the peace movement, a war that would bring the Peace of God—that is, the protection of the Church—to Jerusalem, which was now called "the holy city."

In declaring that the Saracens must be stopped, Urban said that they had "destroyed churches and devastated the kingdom of God." Urban challenged Christendom: "Oh what a disgrace that a race so despicable, degenerate, and enslaved by demons should thus overcome a people endowed with faith in Almighty God and resplendent in the name of Christ. Oh what reproaches will be charged against you by the Lord Himself if you have not helped those who are counted like yourself of the Christian faith."

Urban's speech became for the West what the tenth-century Hamdanid sermons became for the East, a textbook model for rallying the troops. It contains all of the traditional lies by which people are convinced to die and kill.

The enemy is evil—in this case despicable. We, on the other hand, said Urban, have God on our side. It was an Augustinian just war. Those who did not support the war should be and would be singled out as immoral for failing to support the cause—just as in every war those who refused to fight have been vilified by the warmakers. Even questioning a war must be attacked as a sign of suspicious weakness. In June 2005, White House adviser Karl Rove accused the Democrats, because they were questioning the war in Iraq, of wanting to "offer therapy and understanding to our attackers." The fact that no Iraqis had attacked the United States was irrelevant. The point in 2005, as in 1095, was that a failure to hate the enemy, once an enemy had been declared, was unacceptable.

Urban also claimed that the soldiers would be rescuing a poor oppressed people who desperately needed their help. This tactic generally works best if a case can be made that the people in need of being rescued are people like us. This was why Abraham Lincoln preferred to speak of "saving the Union" to "freeing the slaves," why Roosevelt wanted to save freedom rather than save the Jews, and why Ronald Reagan in 1983 did not want to rescue the black Grenadians from an evil coup d'état but instead claimed he was rescuing a handful of American medical students. White Christians generally want to rescue white Christians, which was at the heart of Urban II's message.

Of course not all of these elements are always lies, though they were in this case. The Nazis were actually worse than Allied war propaganda's depiction of them. But history teaches that somewhere behind every war there are always a few lies used as justifications.

———

When Urban II finished his speech, those present shouted, *"Deus volt!"*—It is the will of God. *Deus volt!* became a battle cry for the Crusaders. The Christian version of a "holy war" had been established, and warfare became Christian. Clergy even asserted that a Christian could obtain divine salvation by going to war against the

Saracens. The concept of holy war is one of many ideas that Christians and Muslims borrowed from the Old Testament, which describes numerous wars sanctified by God to deliver God's wrath. In the promotion of the Crusades as a holy war, the Church made frequent references to the Maccabee victory in which the Jews had retaken Jerusalem, the basis of the Chanukah holiday. This was at a time when Jews had little regard for Chanukah as a holiday, but despite Jewish ambivalence this martial festival was the one moment of the Jewish calendar that excited Christians. Holy war also borrows from the Roman wars in which the enemies were considered "barbarians" regardless of the sophistication of their civilizations.

The labels *Crusade* and *Crusader* did not come into usage until one hundred years and four Crusades after Urban II's rallying cry. By then, the concept of a Christian holy war was well entrenched. The First Crusade was termed a pilgrimage. In this Orwellian lexicon of the Church, in which a "peace movement" promoted warfare, why couldn't a ruthless bloodletting be called a pilgrimage? According to Muslim sources, the Christians killed 70,000 people in the taking of Jerusalem. In time the figure grew to 100,000, many of whom were reportedly slaughtered in the Dome of the Rock Mosque, whose siege by Christian soldiers on a holy mission was described by the crusader Raymond of Aguilers:

> Some of our men (and this was more merciful) cut off the heads of their enemies; others shot them with arrows, so that they fell from the towers; others tortured them longer by casting them into flames. Piles of heads, hands, and feet were to be seen in the streets of the city. It was small matters compared to what happened at the Temple of Solomon, a place where religious services are ordinarily chanted. What happened there? If I tell the truth, it will exceed your powers of belief. So let it suffice to say this much at least, that at the temple and portico of Solomon, men rode in blood up to their knees and the bridle reins.

This was not what Augustine had had in mind, and although the papacy made constant subliminal references to Augustine and his concept of just war, the actual phrase "just war" was carefully avoided and rarely was the name of Augustine invoked during the two centuries of Crusades.

At the start of its campaign, the Church manufactured for propaganda purposes both the threat and the evilness of the enemy. The Church had spent many years developing a Western hatred of Muslims so that it could take Muslim lands. This, too, is a lesson: a propaganda machine for hate always has a war waiting. The adversary must first be made into a demon before people will accept the war. This was why during the Cold War, the U.S. government became infuriated at any suggestion that the Soviets were their "moral equivalent." Eleventh-century chroniclers drastically revised early medieval history to demonize Muslims. A famous example is the beautiful eleventh-century poem *Le Chanson de Roland,* which depicts a 778 engagement in the Pyrenees between Charlemagne and the evil Saracens, describing an ambush by the deceitful Muslims in which the Christians valiantly defended themselves. Like the Muslims, the Christians, too, could write poetic war propaganda. In truth, Charlemagne had already made a deal with competing Muslim leaders and easily took Spanish cities by a prearranged collaboration. But then, having not fought any real battles, on his way back to France he sacked Pamplona, a Basque city. It was the Basques and not the Muslims who attacked his rear column in the mountain pass. In reality there were not two all-powerful forces, the Christians and the Muslims, but many warring groups. Urban II unified the Christians by creating this myth of a single, all-powerful Saracen, which for a time was a great Christian advantage since there was no unity among the Muslims.

Another example is the eleventh-century account of the 732 rout of Emir 'Abdarrahman by Charles Martel near Poitiers, France. At the time this was just one more battle in an endless series between various warlords, some of whom were Muslim and some of whom were Christian. In eighth-century Europe, Christians fought Christians, Muslims fought Muslims, and sometimes Christians fought Muslims. But in the eleventh century this battle took on great symbolic importance because it had been the northernmost engagement of the Saracens and they had lost—thus the Saracens had been stopped at Poitiers by Charles Martel. The myth has endured, helping to keep the fires of anti-Muslim hatred fanned in Europe. Edward Gibbon, the eighteenth-century British historian, wrote that if it wasn't for Charles Martel, the Quran might now be taught at Oxford "and her pulpits might demonstrate to a circumcised people the sanctity and truth of the revelations of Mahomet." In modern times Martel has been adopted as the patron saint of the French extreme right.

Eleventh- and twelfth-century Christian leaders were troubled and angered by a tendency of Christian lords to make alliances with Muslims. These could have ruined the entire concept of holy war. In the late ninth century, Pope John VIII had forbidden "impious alliances" with Muslims. His reason was uniquely Catholic in its mysticism. Since all Christians were part of the body of Christ, a Christian who joined with a non-Christian was tearing the limbs off Christ's body. This, too, was power politics. How could the Church control Christian peace and war if there were Christian lords making separate deals with Muslim lords over whom the Church had no power?

When Urban set out to demonize Muslims, he was aided by the fact that the average Christian knew almost nothing about Muslims. There is little historic evidence that the Christians of the Middle East were being oppressed by Muslims. But the Muslims were the enemy. In fact, the word *Saracen* started to be used for anyone who was an enemy. Even other Christians were sometimes cursed as "Saracens."

———

The First Crusade, the only militarily successful one of the six Christian invasions, took the Muslims by surprise. They had been fighting each other and had not expected an attack from Western Europe. Searching for explanations as to why they were being assaulted by a people from another part of the world, Muslims turned to mysticism and astrology, noting that Saturn was in Virgo.

Until then, lesser *jihad* had been the duty of a community but not necessarily all individuals, leaving a choice for those who did not wish to fight. But during the Crusades, Muslim leaders declared that when Islam is attacked, *jihad* is the duty of every individual.

After the Crusades the interpretation of *jihad* became hard line. Ebu's Su'ud wrote in the sixteenth century that peace with infidels was impossible and fighting should be permanent and unending.

The Crusades were about power, not religion. And the Muslims understood this. Initially, they began looking for ties and seeking negotiations with the four new Mediterranean kingdoms the Christians had established in the Middle East. But slowly they built their own war propaganda machine. Just as the Christians established a term for their enemy—the Saracens—the Muslims began calling all the Christian intruders *al-frani,* the Franks. Clerics began teaching that defeat at the hands of the Franks was God's punishment for their failure to carry out their religious duties. And one of those duties was *jihad.* By reviving the culture of *jihad* the Saracens were able to build a counter-Crusade and drive out the Franks. It has happened throughout history: peoples who go to war tend to become mirror images of their enemy—another lesson.

———

In the thirteenth century, Muslims became the enemies of Islam when Mongols, who had converted to Islam, invaded and sacked the Islamic cultural center, Baghdad, in 1258. In the midst of the Mongol disaster, a brilliant young Sunni named Ibn Taymiyah, sometimes known as Shaykh-al-Islam, started writing the first of what was to be 350 works on Islamic law. He completely rejected greater *jihad.* To him *jihad* meant violent warfare, and he insisted that it was the obligation of all fit males to fight. It is Ibn Taymiyah who is quoted today by Osama bin Laden and other "Islamic militants."

The Christian world did have a few voices of moderation. Roger Bacon, the thirteenth-century Franciscan scholar from England, argued that Muslims should be converted rather than killed and that if the Church treated them well they would gladly convert. An Italian theologian of the same period, Thomas Aquinas, argued that the evil done by Muslims did not justify killing them. While these were sincere pleas for nonviolence, they were completely accepting of the false premises with which the violence had been justified. Bacon was agreeing that Islam had to be eliminated, and Aquinas still believed that Muslims necessarily did evil.

Most warmakers try to claim that theirs is a holy war, a just war, that God is on their side, because their cause is just. In the United States the often-repeated inanity, "God bless America," though technically a request, is generally used as a declaration, God blesses America. And war is seldom far behind such assertions—a holy war at that.

It is not surprising that the counter-Crusade and its war cries continue to echo in the Muslim world. Islamic militants from Palestinian Hamas to Libya's Muammar Qaddafi use Crusade and counter-Crusade imagery in speeches to rally the faithful. What is more surprising is that in the West, where the Crusades represent a humanitarian atrocity, an unconscionable act of aggression, a military failure, and one of the worst mistakes in the history of international relations, they also remain a model. Images of the Middle Ages and the Crusades in the movies, video games, and toys by corporations such as Disney steep children at an early age in the culture of warfare and killing. Urban's rallying cry has been copied over and over again. Contemporary right-wing American evangelists such as Billy Graham call their campaigns "crusades." In 2001, when U.S. president George W. Bush announced his "war on terror," his words echoed the messages of Pope Urban II. He even used the word *crusade.* Though George W. Bush may not even have known who Urban II was, Urban's famous speech had become the standard way to sell a war.

IV

TROUBLEMAKERS

How he would have lashed out against anyone who dared to
eat pork on a Friday, and yet now he cannot make the shed-
ding of men's blood a matter of conscience....

—PETR CHELČICKÝ
on his 1420 debate in Prague with
Jakoubek of Stříbro

During the Middle Ages, the customary way to reject militarism was to retreat to a monastery. In fact, the monastic movement was to a large degree a rejection of the Church state with its wealth, power, and wars. This is why the Church refused, and still refuses, to support them. Monks and nuns produced crops, bread, jam, wine, liqueurs, and cheeses while they illuminated stunning manuscripts and sang some of the most beautiful music mankind ever created. Monasteries became citadels of learning in a violent age, enclaves for Christians who refused to take up arms.

Though the Church put aside its interdiction on Christians bearing arms, many of the monastic orders, such as the Franciscans, rigorously maintained the rule. At the same time, many sects arose that operated like the early Christians, outside of the machinery of state, shunning the politics of power. One of the first such groups was the Cathars, in the French-Catalan region of Languedoc.

The Cathars were inspired by the pre-Constantinian Church and traced their theology to a third-century Mesopotamian prophet named Mani, who had contact with not only Christians but Buddhists when he traveled to India. Mani taught a dualistic theology in which there was a realm of God and a realm of Satan, forces of light and forces of darkness. Manichaeism had escaped coopting by a state and so remained a nonviolent theology. This independence, of course, always has a price, and Mani died in a Persian prison.

In subsequent centuries Manichaeism spread. Augustine in his youth had been a Manichaen and later converted to Christianity. For centuries, followers of Mani had taught a message of nonviolence, or at least pacifism, throughout Europe as far north as England, where the religion developed into different sects with an increasingly Christian flavor. In the Balkans, Manichaeism created a sect called the Bogomils, God's Loved Ones. Bogomils were vegetarians who refused to kill animals because of a belief in reincarnation. The killing of humans was forbidden, and since an animal could have a human spirit, thus the killing of all animals was forbidden. But Bogomil pacifism suffered the same fate as Christianity. The religion was adopted by the ruling families of Bosnia, who quickly renounced the ban on killing, and after being conquered by the Turks they converted to Islam.

But the Cathars of southern France were not merely pacifists. They were nonviolent resisters who actively promoted their cause and attacked Roman Catholicism, rejecting the sacraments of the Church, including marriage, because they saw the medieval Church as a fraud, and refused to pay taxes to support the sham. They also scoffed at the notion of private property. The Cathars rejected all killing because it was against the teachings of Christ. They also did not accept the right of the state to kill, either in warfare or in the guise of capital punishment, and consequently they refused to participate in government.

The Church dealt with the Cathars in much the same way as it dealt with the Muslims. First it created myths to demonize them. The name *Cathar*, by which history has known them, was itself a Church-invented pejorative, meaning cat worshipers, because the Church insisted that Cathars kissed the anus of cats. Cathars also were said to eat the ashes of dead babies, hold orgies, even with family members, and practice anal intercourse. They were, according to the Church, strange, perverse, evil heretics. Those who did not want to use the slanderous label called them Albigensians, from the town of Albi, a name rejected by most historians because they came from a much wider area. Almost lost to time was their own name for themselves, "good Christians."

In 1209 the Church unleashed a crusade against the good Christians. It is impossible to say what would have happened had the Cathars held to their beliefs and refused to take up arms. Surely many would have been killed. But what would have been the impact of "soldiers of Christ" slaughtering unarmed good Christians? How long would this have been able to continue as a "holy war"? Instead the Cathars divided into the "pure" and the "fighters," who took up arms. By defending themselves, the Cathars gave an appearance of legitimacy to Pope Innocent II's claim that some wars were just. The campaign took exactly one hundred years, but every last Cathar, both armed and pure, was killed. The moment they engaged in the fight, thereby capitulating to the pope's values, the Cathars had lost.

History teaches over and over again that a conflict between a violent and a nonviolent force is a moral argument. The lesson is that if the nonviolent side can be led to violence, they have lost the argument and they are destroyed.

——

For centuries, unending, almost uninterrupted warfare raged throughout Europe. The priests, noblemen, statesmen, and generals who made the wars filled the pages of history, while those small, persistent, persecuted sects that said, "This is wrong and we will not participate," were slotted for small footnotes. If, as Napoleon was to assert, history is on the side with the biggest artillery, it certainly has little time for those with none, most of whom, for centuries, were powerless commoners.

In 1170, saying he was following the teachings of Christ, a wealthy merchant in Lyons, Pierre Valdes, suddenly gave all his material wealth to poor people, and persuaded others to do the same. Originally the movement was called "the Poor Men of Lyons," but as it spread, the French began referring to its members as Vaudois, which came into English as Waldenses or Waldensians. In the thirteenth century, they seem to have become influenced by the Cathars and, like the Cathars, rejected all killing, including capital punishment and warfare, and refused all military service. They taught that patient suffering was the Christian way to resist and that fighting back was an urge that came from Satan. They stated that the fourth-century alliance between the emperor Constantine and the Church was the origin of a complete corruption of Christianity. Waldensianism spread throughout Western Europe, mainly among the poor, which should not be surprising for a group that required giving up wealth. They survived and became part of the Protestant Reformation.

Another movement, called the Taborites, rose up in the Czech lands of Bohemia and Moravia in the fifteenth century. Inspired by Jan Hus, a Prague priest of peasant origin who, in turn, was inspired by Waldensians, the Taborites also rejected militarism, war, and capital punishment. Hus, burned at the stake for heresy in 1415, was exalted as a martyr for the cause of Czech nationalism. The Taborites were certain that a better world, without violence, was arriving, and they did not have to do a thing to promote it but wait for the second coming of Christ. Some grew impatient with this passive approach but did not seem to know how to turn their pacifism into nonviolent action. And so Jan Želivský, a Taborite leader who often invoked the commandment "Thou shalt not kill," became an advocate of violent revolution. Like the Cathars, the Taborites responded to the threat of a Crusade against them by raising their own army. Like the papacy, they justified this reversal by claiming that they were doing God's work. They became "the Warriors of God," not only defeating the Crusade but terrifying European populations with offensive attacks.

In 1420 it was the Taborites who drove off a Crusade whose real purpose was to seize power in Bohemia and marched into Prague to save it from the invader. Among the holy Taborite warriors who saved Prague was a remarkable man named Petr Chelčický. It is not known why Chelčický participated in the battle for Prague. Some historians believe that he was there only as an observer, others that he had experienced a momentary lapse of faith. If he was a combatant it was a singular aberration in the life of a man who had embraced nonviolence as the only Christian approach long before the fight in Prague and continued to do so long after. While in Prague he had sought out the noted theologian Jakoubek of Stříbro. Jakoubek believed that although warfare was unchristian, Christians could engage in it for a "just war"—that is, if the cause was in the service of God. According to Chelčický's writings, when he challenged Jakoubek to show one passage anywhere in the New Testament that said that it was permissible for a Christian to engage in warfare, Jakoubek was forced to concede he had no such authority but was relying merely on the teachings of "saints of old."

Chelčický urged Christians to return to the teachings of Christ, "the law of love," as expressed in the New Testament. His principal work, *The Net of Faith,* called for just such a Christian renewal. He believed that it was not possible to kill, regardless of the circumstances, and still practice what Jesus called love. Government, as it was known, was intrinsically pagan, for it could not operate under Christ's law of love.

Significantly, Chelčický, like many Hussites, Waldensists, and Taborites of the period, erroneously believed that Pierre Valdes had lived in the time of Constantine. According to the legend, when Valdes's good friend Pope Sylvestre allied the Church with Constantine, betraying and undermining Christianity, Valdes had led his followers into the wilderness to live a pure life uncorrupted by the polluted Church.

Chelčický recognized, as did Paul the Apostle, that government was necessary to keep order among those who did not live under the Christian law of love, but this was not to be the work of Christians. This meant that Christianity was to be a marginal and probably persecuted group. That, according to Chelčický, was the proper role of Christianity. "There can be no power without cruelty," he wrote. "If power forgives, it prepares its own destruction, because none will fear it when they see that it uses love and not the force before which one trembles."

Chelčický was one of the first to see that the cause of perpetual war lies not in the nature of man but in the nature of power. To establish a world living in peace would require the abandonment of power politics, both on the grand scale—states trying to bend other states to their will—and on the small scale—legal systems coercing social behavior with the threat of prison.

Chelčický's writings also explore what today would be seen as a nearly Marxist analysis of society, though, ironically, in the 1950s, official Czech communism rejected him as "petit bourgeois." According to Chelčický, the few accumulate wealth by exploiting the labor of the impoverished many. He also saw war as a conspiracy in which the poor were duped into fighting to defend the privileges of the rich. If all poor people refused to fight, he argued, the rich would have no army and there would be no war. He was even opposed to universities promoting a militaristic, warmongering, wealth-hording society—a point made again by many students in the 1960s.

Around 1460, about his seventieth year, Chelčický died. But his teachings continued to find adherents. Calling themselves the Unity of Brethren, twenty years after Chelčický's death they had ten thousand members among the Czechs. Forty years later a similar pacifist movement, the Anabaptists, rose up in German-speaking Switzerland. Such movements became part of the Protestant Reformation. Martin Luther insisted on nonviolence in personal relations, though he accepted warfare. The Anabaptists of Zurich broke with the Reformation in 1525, when Reformation leader Ulrich Zwingli would not go as far as they wanted on such issues as a complete ban on violence.

Between 1525 and 1800 more than two hundred decrees were issued by various European governments denouncing the Anabaptists. These denunciations were often based on the Anabaptist position that conscious adult believers should be baptized and not unknowing infants. But they were also denounced for their refusal to bear arms. Like the early Christians, their refusal extended to any participation in the state. They also refused to swear oaths, including oaths of allegiance, a stand that can be traced at least as far back as the Cathars. In the Middle Ages oaths had been essential to the warrior code, and rejecting them was rejecting war. Today oaths seem less important, and most contemporaries never swear an oath. But Americans are still required as young children to pledge allegiance to the flag, one of the first steps in conditioning young Americans for war. This assertion can be easily tested: denounce the saying of the pledge of allegiance and see if the people who are outraged are not the same people who promote war.

The Anabaptists, seen as a threat to the state, were driven out of one town after another, exiled, and sometimes executed. Despite this, or perhaps as a result of their repeated exile, the movement spread as far west as Alsace, throughout Germany, Austria, and the Tirol. They were particularly disliked in Hapsburg lands, as they refused to fight the Turks. One Anabaptist leader, Michael Sattler, argued at his 1527 trial that the Turks were non-Christians and knew nothing of Christ's teachings, while the Christians who would go to war against them were "Turks after the spirit," pagans who had rejected Christ's teachings. Sattler said he would rather battle them than the Turks. He was executed. Of all the dangerous Anabaptist heresies, none was more threatening to the state than this refusal to fight Turks.

But the movement continued to spread. By the seventeenth century it had traveled north to Holland and east to Poland. In 1658 the Polish Catholic Church forced Anabaptists to either convert to Catholicism or leave. Many left—for Transylvania, Germany, and Holland.

In 1572 Holland began its long war of independence against Spain. Suddenly groups appeared—the Mennonites, the Waterlanders—refusing military service and stating that they would not fight the Spanish. The Mennonites are so called because they followed Simon Menno, a sixteenth-century Dutch Catholic priest who had left the Church to work with the Anabaptist movement. In 1572 the Mennonites went to the monarch, William of Orange, challenging the notion always hurled at pacifists, that they were unpatriotic. They told William that they wanted to find a nonviolent way to support his cause. The Mennonites offered to raise money for the financially strapped king and with dazzling speed raised a sizable sum from their members. It became a tradition in Holland to tax Mennonites in time of war in exchange for exemption from military service. In the seventeenth century they agreed to serve in noncombatant roles. When Louis XIV of France invaded Holland in 1672, Mennonites volunteered as firemen in besieged cities. Such alternative efforts make pacifists better accepted. But the troubling issue of whether offering services is helping the war would remain for centuries.

A good lesson for the military is that the longer wars last, the less popular they become. So it is not surprising that a war known as the Thirty Years War would be deeply unpopular. Of course no one planned for the war to last that long, and the name came later. Wars usually start well for the war promoters. The Urban II speech is made; the enemy is declared despicable; God's stand, firmly behind the cause, is declared; and the killing is nothing short of the honorable and patriotic thing to do. In the case of the Thirty Years War no one could find as tidy a justification as "driving out the infidel," but the cause was outlined with the prerequisite simplicity nevertheless. The war was basically about the dismantling of the Holy Roman Empire, which, as every schoolchild learns, was never holy, was German, not Roman, and was never exactly an empire.

Today it is difficult to explain what those two generations of combat were about—neither the first nor last war to lose meaning in hindsight—yet European monarchs were able to raise huge armies nevertheless to slaughter each other from 1618 to 1648, for the cause, or causes. It was as though World War I had continued into World War II without a break in the fighting.

Most of the fighting took place in Germany, decimating the German population, destroying German agriculture and trade, and giving German peasantry a jaundiced view of war for the next two centuries, during which time German pacifist and antiwar movements flourished. Numerous groups, such as the Spiritualists, mystics who believed inspiration from the Holy Ghost was more important than Scripture, opposed war. The year the war ended, 1648, a mystic named Paul Felgenhauer wrote *Perspective of War,* a book asserting that the recent calamity in Germany was the beginning of the end of the world. He argued that a complete rejection of warfare was the only reasonable stance. Other books rejecting warfare, even defensive war, were written after the war by other mystics, such as Christian Hohburg of Hamburg and Annecken Hoogwand, who declared war a sin. Pietism, a sect that challenged the Lutheran establishment much the way earlier sects had challenged Catholicism, emerged with a strong antiwar message. From this movement came the German Baptist Brethren, popularly known as the Dunkers, who embraced a style of pacifism similar to that of the Mennonites.

Enduring a civil war followed by a revolution, seventeenth-century England also experienced enough warfare to stimulate antiwar movements. In this setting, Quakerism, a mystic religion that was neither Catholic not Protestant, rejected both sides in the English upheaval. To a Quaker, Oliver Cromwell, who led the Puritans to power through his military prowess, was the Constantine of Puritanism. By establishing the religion he had destroyed its principles.

The Quakers were so named, mockingly, after the physical habits of its founder, George Fox, when in a state of spiritual possession. Though initially persecuted, especially by local government, the Quakers were largely protected by Cromwell as one of theirs, once the Puritans came to power. But Quakers provoked the political establishment by refusing the taking of oaths and the tipping of a hat as a sign of respect. Only gradually did they become pacifists, and once they adopted an uncompromising antiwar stance, their persecution, including prison, public beatings, and whippings, became widespread. In the 1670s their meeting houses were forcibly closed, even physically dismantled.

Spin-offs of Puritanism also rejected violence, such as the Diggers, a short-lived movement that tried to establish an egalitarian commune in rural Surrey in 1649. They were destroyed by a violent mob, but were much talked about and imitated by Yippies and other 1960s movements in the United States. Gerrard Winstanley, the founder of the Diggers, had called war "a plague" and wrote, "We abhor fighting for freedom." This seems a strange paradox at first until one reflects on how often in history war is justified as a fight for freedom and how rarely that is the true goal.

War burdens the working class, and that was traditionally the source of antiwar sentiments. Quaker founder George Fox was a shoemaker. But William Penn, a convert to Quakerism in 1667, at the age of twenty-three, was the son of a British admiral and an aristocrat with a personal acquaintanceship with King James II. Even after his conversion, he was reluctant at first to drop the aristocratic fashion of wearing a sword. Penn wrote of the persuasive power of love and the unchristian, warlike nature of Christians, whom he termed—in the ultimate seventeenth-century European insult—to be worse than the Turks. It was Penn who offered the simplest formula for ending war, that it starts with an individual refusing to fight. According to Penn, "Somebody must begin it."

V

THE DILEMMA OF UNNATURAL PEOPLE

The governments of the earth have built up a structure that exists only by the power of money. The head of the land—the Queen—is honored in proportion to the pomps and vanities of her immediate attendants. Her governors all hold out their hands for their wages, without which their patriotism would shrivel up.

—TE WHITI, Maori chief, 1879

A not insignificant piece of misinformation passed on to American schoolchildren is that Pennsylvania was named after the Quaker leader, William Penn. Penn was the founder of the colony after obtaining a charter from King Charles II of England for a "holy experiment." But Charles named the colony not for this questionable Quaker pacifist but for his father, the great British admiral Sir William Penn, who had served in the First and Second Dutch Wars and captured the island of Jamaica from the Spanish. This was the kind of man kings named holdings after—not his son, who while the admiral was fighting for England was expelled from Oxford for unorthodox religious beliefs and had been getting into trouble ever since.

Quakers went to other colonies as well. The first recorded case of an American conscientious objector was Richard Keene, a Quaker convert who refused training in the Maryland militia and was fined and angrily threatened with a drawn saber by his commanding officer. The first two Quakers known to arrive in America— although a missionary also arrived in Maryland around the same time and two years earlier a Long Island resident had converted during a visit to England—were two women missionaries who landed in Boston in 1656. Within two months they were expelled. Seven more arrived and Massachusetts and Plymouth colonies both began passing increasingly brutal anti-Quaker ordinances—from heavy fines to men and women being stripped to the waist and beaten, to mutilations of ears and tongues. By 1658, four Quakers had been executed.

But Pennsylvania was different. Penn's holy experiment attracted to America not only Quakers but Mennonites, Dunkers, and other pacifists and idealists. Penn had intended Pennsylvania to be a model for the world, a pacifist state, what Penn called "a precedent." The early years of the colony were marked by an unusually open relationship with the Indians and a firm stance against war.

Pennsylvania militias were volunteer forces. The colony did not accept British conscription in local militias as the other colonies did. Although in Rhode Island Quakers were automatically exempt from military service, most other colonies insisted that conscientious objectors pay a fine or hire someone to serve as a replacement, neither of which was an acceptable alternative to Quakers. In time of war this led to persecution, often imprisonment. Even before the French and Indian War, an American extension of the nearly global Seven Years War, there was almost constant warring with Indians and between European powers in North America and the Caribbean, including King William's War (1689–97), Queen Anne's War (1702–13), and King George's War (1744). To live in a European colony was to constantly be called upon to fight Europe's wars.

During all these conflicts there were small numbers of Americans, not all of them religious, who refused to fight. In 1675 a few men refused to participate in preparations to defend against an Indian assault on Boston. In the early eighteenth century the Massachusetts colony found it necessary to pass a law establishing prison sentences for those who refused to bear arms. Indian attacks were said by some to be God's punishment for the colony allowing the presence of Quakers.

———

Historians commonly suggest that the acts of political leaders are the result of the slow dissemination of—and society's osmotic absorption of—the ideas of intellectuals. While it is true that political leaders absorb ideas from the academy, they also requisition ideas, embracing those thinkers who provide them with the intellectual underpinnings for what they want to do. In seventeenth-century Europe, as policies of warfare, colonialism, and slavery were expanded, a great deal of thought went into rationalizing these acts. No nation of the period was as successful at providing intellectual justifications for these policies as England. Richard Tuck, a professor of government at Harvard University, has stated that the reason England was the most successful European colonizer is that it had the best intellectual underpinnings for this role.

Alberico Gentili, an Italian who in the late sixteenth century became an Oxford professor of civil law, expanded the concept of defensive warfare and established the principle, recently touted by the George W. Bush presidency to justify invading Iraq, of the preemptive strike. "No one ought to wait to be struck unless he is a fool," Gentili wrote. The argument was not new. The Romans justified attacking Carthage by claiming that Carthage would at some point in the future attack them. But given the state of intra-European relations in Gentili's time, there was not a moment in the next few centuries when some European power could not reasonably fear the military of another and unleash a preemptive strike, which is why that era saw scarcely a moment without warfare. "For certainly," in Gentili's words, "as long as men are men, the sons of Prometheus and not Epimetheus, and as long as reason is reason, a just fear will be a just cause of a preventative war."

Gentili went even further in his Oxford lectures, stating that being the aggressor was also justified in the defense of property and holdings. He specifically justified the Spanish conquests in the New World, neatly piecing together references to the Romans spreading liberty with references to Augustine's just war to preserve the values of society and Urban's demonization of the infidel. Not surprising for a new age of imperialism, the original imperialists, the Romans, were being cited more and more. The Indians, like the Gauls, *deserved* to be conquered.

> The cause of the Spaniards is just when they make war upon the Indians, who practiced abominable lewdness, even with beasts, and who ate human flesh, slaying men for that purpose. For such sins are contrary to human nature. . . .

Gentili employed a similar logic to justify slavery, underscoring the justness of enslaving the "wicked." This was an ancient argument. Aristotle wrote about those who deserved to be enslaved. But here it was applied to wicked races, races that deserved to be conquered and deserved to be enslaved. The slavery of the sixteenth through the nineteenth centuries was necessarily racist because it argued that Africans should be enslaved because of their inherent moral inferiority. Thomas Jefferson, who professed to be opposed to slavery as a "political and moral evil" even while owning slaves, believed that while all men were created equal, black men were less equal. In his *Notes on Virginia,* he amply discusses the inherent inferiority of African people. Since the inherent inferiority of other people could be invoked to justify both warfare and slavery, it was not a coincidence that the promoters of warfare and of slavery were often the same people, and that abolitionism and nonviolence, often, though not always, went hand in hand. In 1783, when the U.S. Constitution was being written, the Quakers were the loudest voice for abolishing slavery.

In the seventeenth century the English justification for war and colonialism was further enriched by another Oxford scholar and member of Parliament, John Selden, with his conclusion that "extending empire was a good enough reason" to go to war, because man had a natural right to acquisitiveness. This thinking was further developed by another seventeenth-century Oxford thinker, Thomas Hobbes. Though Hobbes believed in monarchy as the most efficient form of government, he strongly influenced the American Founding Fathers with his belief that the origin of power was the people, who only submitted to a sovereign because they required protection. Yet Hobbes, like Selden, believed that war and violence were part of the natural order. In his central thesis on government, *Leviathan,* he wrote that man had a selfish nature and that continual warfare was his natural state. A century later, the French philosopher Jean-Jacques Rousseau argued that beyond being the individual's natural state, warfare was the natural condition of the nation-state. Hobbes also believed, like Selden, in man's acquisitive nature, and that until contracts to the contrary were established, he had the right to take what he wanted. Hobbes had worked for Lord Cavendish, one of the leaders of the Virginia Company, which invested in North American colonies, and had been awarded a share in that organization's American holdings.

John Locke, another Oxford intellectual who greatly influenced

America's Founding Fathers, was a firm believer in the middle class and its right to property and was a strong enthusiast for British colonialism. He was directly involved in the establishment of Carolina, where an island, now called Edisto, was for a time named after him. He also invested in the Bahamas and the Royal Africa Company. Locke believed that God gave man land to use and enjoy and that since Europeans with their advanced agriculture used it better than "savages," they had the right to take it by force. According to Locke, Europeans had the right to punish others for not living by what he deemed "natural laws." He also believed they had the right to take slaves, over whom they were free to exercise the power of life and death.

———

These were the radical thinkers of the day who were shaping revolutionary thought in America. But there were other, very different, kinds of radicals in America, such as William Penn. Pennsylvania—and eventually neighboring colonies—drew people who denied the state its Hobbesian rights to war, colonial expansion, and slavery. In Penn's "holy experiment," the Quakers controlled the Assembly and made the rules favorable to the nonviolent sects. These sects, which had shunned political participation in Europe, now voted and actively worked to keep the Quakers in power. Many, such as the Mennonites, for the first time were living under a type of government in which they could fully participate. The colony assigned land on the western frontier to the warlike settlers—people not from peace sects—whose implicit role was to confront the Indians; but the frontiersmen resented the pacifists, who were given more secure eastern lands that often also happened to have richer soil. A tension developed in Pennsylvania between eastern establishment and western frontiersmen that was a microcosm for what would be seen in the country as a whole after independence.

The Quakers ran their colony of Pennsylvania as though it were an independent state, adopting a foreign policy completely out of line with the British Empire. Not only did Pennsylvania refuse to conscript militias to fight the French, they would not fight Indians and independently negotiated peaceful and friendly relations with them.

Had Quakers controlled all of the colonial legislatures and not just that of Pennsylvania, the history of North America—and perhaps, by example, all of the Americas, Africa, and much of Asia— might have been different. The Quakers did not believe that non-Christian people were unnatural and needed to be conquered. In North America they not only tried to teach Quakerism to the Indians by example, they also directly preached it to them. They had little success. The Indians were not just dealing with Quakers, they were caught between two ruthless European empires, both of which coveted their land. As long as the British and the French were active in the New World, nonviolence made little sense to these people.

Samuel Bownas, a British Quaker who soon after he landed in Maryland began debating important colonial figures on Quaker issues, very quickly found himself in colonial prison. In his autobiography he described a conversation with four Indians while he was an inmate in a Long Island prison in 1702. He explained that while most white men believed that killing their enemies was acceptable behavior, Quakers believed that it was wrong, that Quakers "rather endeavor to overcome our enemies with courteous and friendly offices and kindness, and to assuage their wrath by mildness and persuasion." The Indians agreed that "this was good. But who can do it?" The leader of the group argued, "When my enemies seek my life, how can I do other than use my endeavor to destroy them in my own defense?" The four agreed that if everyone adopted this point of view, "there would be no more need of war, nor killing one the other to enlarge their kingdom, nor one nation want to overcome another," but they did not think many people would take up this belief. Bownas gave the standard Quaker response: "All things have their beginnings." The Indians agreed that if Bownas had his way, "things will go well" in the world.

The frontiersmen of the western border of Pennsylvania did not follow the dictates of the Quakers. An increasing number of atrocities were committed throughout the colony by frontiersmen against the Indians, notably a massacre in Lancaster in 1763. In 1764 some 1,500 western settlers marched to Philadelphia to protest the refusal of the Quaker-controlled Assembly to pay bounties for Indian scalps. There was growing disenchantment with the Quaker Assembly, and it was only because the pacifist sects voted in a block that they were able to stay in power long after they had lost the support of the majority of Pennsylvanians. Finally, in 1756, they were voted out of power, and the new assembly dropped their nonviolent stance. Mennonites and other nonviolent sects retreated to their traditional posture of nonparticipation in government and were seldom heard from again until the fight over slavery heated up in the mid-nineteenth century. Quaker control of the colony lasted only seventy-four years. The central problem was that the pacifist state was part of a larger colonial system that vehemently rejected nonviolence.

———

In the vast history of European colonialism, there are few incidents of nonviolent resistance by indigenous people, leaving unanswered the question of whether this would have worked. What is answerable is that nothing they did try worked. The indigenous people of five continents were facing an intractable enemy from a sixth continent that was convinced that they had the right to steal the land on other continents and destroy the inhabitants as peoples and cultures, and, in fact, that this was the proper thing to do. The Europeans had not only the public and the clergy, but the intelligentsia, the thinkers and philosophers, backing up their program of genocide.

What were these indigenous peoples to do? Most decided to resist militarily, and it can now be seen that this was a disaster. But right from the start it was clear that pacifism was also a route to annihilation. Bartolomé de las Casas, who was born in Spain in 1474, and was possibly the first Catholic priest ordained in the Americas, wanted to establish towns in the Americas where Indians and Spaniards could live together in peace. "For this is nothing else than making the coming and passion of Christ useless ... as long as innumerable human beings are slaughtered in a war waged on the pretext of preaching the gospel and spreading religion." But he could find little Spanish support for such a project.

In 1542 de las Casas claimed to have seen the indigenous population of the Caribbean island of Hispaniola reduced from three million to two hundred survivors. The reason, he said, that this extermination was possible was that "of all the infinite universe of humanity, these people are the most guileless, the most devoid of wickedness and duplicity, the most obedient and faithful to their native masters and to the Spanish Christians whom they serve. . . . Yet into this sheepfold, into the land of meek outcasts there came some Spaniards who immediately behaved like ravening wild beasts. . . ." As Gandhi would observe centuries later, it would take more than meekness to survive European empire builders.

In North America, the Cherokee took a different approach. The largest pre-European nation in what was to become the United States, the Cherokee had a rich culture and a developed agricultural economy by the time Europeans arrived. After seeing that a military response was disastrous, in 1820 they adopted an American-style democracy, with an elected leader, a House, and a Senate. Seven years later they declared themselves to be again a nation, with a capital in Georgia. They developed written characters for their language and wrote laws and published a newspaper. For a few years they seemed able to coexist with their sister democracy, the United States. This might have continued were it not for the discovery of gold on their land.

In 1830 the U.S. Congress debated the "Indian Removal Act." Tennessee congressman Davy Crockett staked and lost his political future on opposing the bill, and when it passed, he left Washington for Texas, saying, "I would sooner be honestly damned than hypo-critically immortalized." The Cherokee went to court and won in the Supreme Court when Chief Justice John Marshall ruled that the Cherokee nation was sovereign and the removal act was illegal. Such an act would have to be negotiated between the Cherokee and the U.S. government.

This would have been a great triumph for nonviolence and the rule of law, except that President Andrew Jackson, the great advocate of Indian removal, was able to find a small Cherokee faction of fewer than 500 people in a nation of 17,000 who were agreeable to removal. A treaty was signed with them and was ratified in the U.S. Senate, passing by only one vote over the vociferous objection of such leading figures as Daniel Webster and Henry Clay. When seven thousand army troops were ordered to force-march the Cherokee, all 17,000, to what is now Oklahoma, the commanding officer, John Wool, resigned in protest and was replaced by General Winfield Scott, whose march, known in history as "the Trail of Tears," *nunna daul Tsuny* in the Cherokee language, was so brutal that 4,000 Cherokees died. The Cherokee had lost their faith in nonviolence and put to death the leaders of the small contingent that had signed the removal treaty.

———

But in a remote corner of the far-flung British Empire, there was a people who took on the British with classic nonviolent activism. The far South Pacific seems to have been one of the last places settled by humans. The few islands beyond Australia are not near anyone else nor on the way to anywhere. The first people to go there were Polynesians. In 950, a Polynesian navigator named Kupe discovered the islands, which are today New Zealand. Following his discovery there was a sizable migration of Polynesians to these uninhabited islands.

Polynesians are usually portrayed in movies as a gentle, peaceful people. But this was often not the case, and certainly not the case with the ones who settled New Zealand and called themselves the Maori, which means "Children of Heaven." The Children of Heaven were armed with flat stone clubs that they wielded with deadly force from a thong attached to the handle. Competing Maori groups fought each other for control of the limited land. Defeated enemies were often enslaved and on rare occasions were eaten.

The Maori resisted the British militarily and from 1845 to 1872 Europeans fought to destroy them and take over their land. A memorial to the soldiers of that war in the Auckland War Museum, a museum dedicated to the glorification of war, says, "Through war they won the peace we know." If they took out the word *peace* and replaced it with the word *land,* they would have had the truth. Toward the end of the war the government was paying bounties for heads, £5 for a regular head and £10 for a chief's. To collect the bounty the killers would carry the sack of heads to the commanding officer's tent where they would dump it out, the heads rolling across the floor. White soldiers and pro-government Maori competed for heads; in one famous incident a Maori shot an insurgent and a white soldier ran up to the wounded man, still alive, and decapitated him to rob the Maori of the £5.

In 1867, with the territory almost entirely under British control, and only about 40,000 Maori still alive in a land of half a million Europeans, a visionary Maori leader named Te Whiti emerged on the southern coast of New Zealand's northern island, in a place called Parihaka. As government troops hunted down the last of the resisters, some fled to Parihaka, where they were allowed to stay, on condition that their weapons were destroyed. The population of Parihaka was growing. In 1869 Te Whiti declared that this was to be "the Year of Trampling Underfoot." The whites took this to mean their final victory, but Te Whiti was actually saying that 1869 was to be the year in which the people of power were to be humbled. He gave the *pakeha,* as the Maori called white people, the bad news: without using violence or force, it was his intention to negotiate a separate treaty between the *pakeha* and the Maori of his district, a treaty between equals. He said that lion and lamb, hawk and wren, cat and mouse, and the *pakeha* and the Maori would lie down together in peace. But the management of the land would remain in the hands of its owners, the Maori. White settlers could remain on the land they occupied and more could come if they wanted. But the Maori would remain the proprietors and no parcels were to be sold off.

His intention in claiming the land peacefully was for the Maori fighters to give up their armed struggle. Many did, flocking to Parihaka and turning in their arms. White people came, too, and Te Whiti received them graciously. They marveled at the elegance and refinement of this broad-shouldered man with strong features and a premature white beard. Paying him what they considered the ultimate compliment, some said that he seemed like a white man.

Like the American colonial revolutionaries before him and Gandhi after him, Te Whiti urged his people to make or grow everything they used and not buy British goods. The Maoris who came from all over New Zealand to hear Te Whiti stayed to grow food. They came with ploughs—more and more ploughs. For years white townspeople noticed Maori headed to Parihaka with ploughs.

Finally, on May 26, 1879, ten years after Te Whiti's announcement, white farmers looked out their windows and saw why the Maori had been collecting ploughs. All over the district, white-held land—land that had not been disputed for decades since its Maori inhabitants had been killed or driven out—was being ploughed by Maori. When the white farmers would run up to them and shout, the Maori ploughmen, always unarmed, would remain calm and polite, and would continue ploughing acre after acre from sunrise to sunset every day. Sir Hercules Robinson, the governor, rode out to see and, according to witnesses, "almost exploded with indignation."

While the government hesitated, a group of angry white settlers declared an independent republic and pledged to kill the ploughmen. Te Whiti instructed the ploughmen: "Go put your hands to the plough. Look not back. If any come with guns or with swords, be not afraid. If they smite you, smite not in return. If they rend you, be not discouraged. Another will take up the good work."

The settlers strangely panicked in the face of this nonresistance. Some built fortresses around their homes. Some dug trenches. The local newspaper called for a "war of extermination." Te Whiti told his followers not to disturb the property of fleeing whites. "If any man molests me," he said, "I will talk with my weapon—the tongue."

The government began arresting the ploughmen, who did not resist. Te Whiti sent out five each day. Hundreds were being arrested and shipped to the South Island, where they were held without trial. Some died from the poor conditions. Still, more and more white-held land was being marked by the furrows of Maori ploughs. The army moved in, building blockhouses and digging trenches as though preparing for a major battle. The ploughing continued. The Maori warriors, who had defeated armed columns twelve years earlier, met the invaders and gave them gifts.

By 1880 the government estimated that it had already spent close to £1 million trying to put down the nonviolent uprising. The Parihakas, identifying themselves by the wearing of a white feather, were becoming an obsession of the press. There was, however, almost no coverage of the fate of some four hundred Parihaka prisoners until a few of them were released. Once their stories were told, their treatment fast became a national scandal and all four hundred prisoners were released.

Hundreds of Maori were out working their reclaimed fields every day. But the local constabulary had raised an army of 2,500. When they attacked Parihakas they were met by young girls singing songs. Though cursed at and threatened with swords and charging horses, they continued singing and playing. When the army finally made its way to the center of town, the entire village was seated on the ground waiting for them. The commander gave them one hour to disperse, but several hours later the entire village was still ignoring their presence. Finally Te Whiti was arrested and taken away. The chief smiled as he walked away under guard. Even with Te Whiti in prison, and later in exile in the south, with troops destroying their homes, the Parihakas still continued to find means of nonviolent protest, such as refusing to pay taxes. In 1897 they began another ploughing campaign.

Te Whiti died in November 1907 and was buried with a cloud of white feathers. The Maori did not take back the land, and the half-million Maori today are only 20 percent of New Zealand's population. But Te Whiti and his movement in Parihaka are credited with stopping a war of genocide that would have meant the end of the

Maori people. What might have been the fate of the Maori with more Te Whitis? What might the Spanish and French have done in the face of nonviolent resistance on Hispaniola? What if there had been a Te Whiti among the Cherokee or the Iroquois? But such leaders are rare, and, as the Quakers said, there has to be someone to begin it.

VI

NATURAL REVOLUTION

We abhor fighting for Freedom. Freedom gotten by the sword is an established bondage to some part or other of the creation. Victory that is gotten by the sword is a victory that slaves get one over another.

—GERRARD WINSTANLEY,
leader of the Diggers, 1650

Someone who had not studied history might have imagined this war-torn world welcoming one peaceful state. But an age-old lesson was seen once again in colonial Pennsylvania. The new Americans could not conceive of power without force. Pennsylvania, they believed, would not be significant, would not have a role to play in the shaping of the New World, if it could not engage in warfare. Many Pennsylvanians who wanted to see their colony at the center of events, among them Benjamin Franklin, actively campaigned to stop support for the Quakers. Franklin's 1747 pamphlet *The Plain Truth* argued against the Quaker policy of refusing to support the British wars against the French and Indians.

The revolutionary movement of the 1770s was well aware that there were those in their midst who were anti-British but opposed a shooting war. In fact, 80,000 Americans belonged to nonviolent sects that would not accept warfare, far more people than the number of troops Washington ever had under his command at any one time. On July 18, 1775, the Continental Congress, meeting in Pennsylvania, called on religious pacifists to contribute in nonviolent ways. The Congress was inundated with petitions from religious conscientious-objector groups as well as from those who denounced these groups as "enemies of Liberty." Both sides finally agreed that those whose religion opposed violence would not be penalized for refusing service as long as they paid taxes to support the war, including an annual fine of two pounds and ten shillings.

The majority of the Pennsylvania Assembly was anti-British but viewed going to war as an unnecessary extreme. They believed that their differences with the British could be negotiated. It was this attitude that made the revolutionaries greatly distrust the Assembly. The revolutionaries also remembered that in the past, Quakers had in some cases successfully converted soldiers to nonviolence and these individuals would no longer fight. The spreading of nonviolence outside the religious sects was reason enough to regard these people as a threat, and they were persecuted in most of the colonies.

In the years leading up to the American Revolution, the radical revolutionaries, those who wanted to break away from Britain and were prepared to go to war, were a minority, but they were the most vocal and articulate and the best organized faction. Proponents of nonviolence know that it is often not the largest but the best organized and most articulate group that prevails. It is not clear that the decision to go to war against the British was the majority opinion in most of the revolting colonies, but the radicals proceeded and made it a fait accompli.

Another enduring lesson of history is that it is always easier to promote war than peace, easier to end the peace than end the war, because peace is fragile and war is durable. Once the first shots are fired, those who oppose the war are simply branded as traitors. All debate ends once the first shots are fired, so firing shots is always an effective way to end the debate. The silence may not last for long, as the War of 1812, World War I, Vietnam, and Iraq, all unpopular wars, demonstrate, but there is always a moment of enforced silence when debate and criticism are banished and this moment gives the war boosters at least a temporary advantage.

In February 1775 the British sent 240 soldiers to Salem, Massachusetts, to seize ammunition and weapons that the rebels were amassing. Though the nonviolent defense of a weapons cache does not truly qualify as nonviolence, the townspeople's plan averted violence and prevented the opening of a shooting war. They simply pulled up the drawbridge into town and made the British negotiate entry, which the British did by giving assurances that they would not disturb the town. Apparently the colonists at the drawbridge were less concerned about the fate of the weapons than the principle that the British army had to ask permission before entering their town. According to Hobbesian logic, such happy solutions only put off the inevitable, which came on April 19, when another British column attempted to seize another rebel arms cache, this time in Concord. Whether or not this qualified as what Hobbes termed Natural Law, the reality was that elements among the rebel movement had decided that they wanted a shooting war, and once that kind of decision is made, it is, as a rule, almost impossible to avoid it. American revolutionaries intercepted the British column in Lexington. The rebels only exchanged a few shots and a number of them were killed. Each side claimed the other side had fired first, though all the casualties of this brief first engagement were on the rebel side. The British marched on to the supply depot in Concord. But the shots had been fired, the war begun, and the debate ended.

Curiously, up until those few shots were fired in Lexington, the rebels, even while arguing for war, had been spectacularly successful at what could be considered nonviolent resistance. Both demonstrating and rioting for a wide range of causes were commonplace in eighteenth-century America. One historian, Paul A. Gilje, counted 150 riots and street actions in the thirteen colonies just between 1765 and 1769. Though rules of class conduct were not rigid, generally the upper classes wrote pamphlets and negotiated, while the lower classes took to the street. The lower classes would cart around effigies of officials at their demonstrations before hanging, burning, or beheading them. Even before television there was a belief that effective nonviolence needed to be visual, needed a sense of theater to attract an audience. When the British passed the Stamp Act in 1765, the colonists staged a series of demonstrations throughout the colonies. In Charleston, South Carolina, two thousand demonstrators protested taxes by burning effigies and then staging a mock funeral for the death of "American Liberty." The stamp officials were forced to resign in every colony but Georgia. The demonstrations were accompanied by a boycott of British goods. The result of all this was that within a year the act was repealed. But the following year the British attempted another taxation scheme, the Townsend Acts, which, because they only taxed imports indirectly, the British hoped would be more palatable.

The working poor were angry about their economic plight and they were not always nonviolent. They attacked and destroyed homes of officials, and looting was not uncommon. The intellectual leaders, being largely men of property, opposed these acts of destruction and tried to keep the street protests orderly. There was clearly a class division, and the upper-class leaders had to negotiate with the street leaders. The former tried to keep elements that they thought of as rowdy out of demonstrations. They sometimes banned black people from participating in demonstrations, convinced that they were an inherently unruly race.

In 1768 the Massachusetts Assembly dissolved rather than collect the Townsend duties. Not entirely nonviolent, the revolutionaries formed mobs to harass customs officials. On March 5, 1770, boys began throwing snowballs at British troops in Boston. The troops began pushing. Men came to the aid of boys. When one British soldier was struck with a club, he responded by firing into the crowd. Other soldiers also fired and five colonists were killed. When the British soldiers were brought to trial, John Adams, a moderate, defended them and noted in defense of the troops that black people were in the crowd. As a matter of fact, a mulatto man, Crispus Attucks, was among the victims. The British were acquitted.

By 1770 the British recognized the Townsend Acts to be another political and financial disaster and repealed them. But the tax on tea remained. This led to the most famous act of nonviolence in the American colonial period.

The American revolutionaries, in their prewar days, were particularly effective in their use of an important nonviolent tool, the boycott. Women began weaving cloth by hand rather than buy fabric from British mills. Homespun became the fashion. Spinning bees became patriotic gatherings. One result of the tea boycott was that Americans very quickly became coffee drinkers. But there were many debates in Boston on how to take the tea boycott even further. On December 16, 1773, sixty revolutionaries, dressed as Mohawk Indians, boarded three ships in Boston Harbor and dumped 342 chests of tea valued at £10,000 into the sea. This was a perfectly managed act of nonviolent protest. There were no incidents of looting or vandalism. According to legend, one padlock was broken and the revolutionaries replaced it.

Though far less famous today than the Boston Tea Party, the crowning achievement of American colonial civil disobedience, the one that John Adams considered the turning point of the American Revolution, came in 1774, before any shots were fired. The colonies were becoming ungovernable and unprofitable. The British were responding with repression, including the so-called Coercive Acts, which cost them more money and tied up more troops. From the point of view of the rebels, the British response was ideal, as it was mobilizing public opinion against England. One of the new repressive measures enacted by the British Parliament, intended as a response to the Boston Tea Party, was the Massachusetts Government Act passed in the spring of 1774. It removed the right of Massachusetts' elected representatives to have a say in the appointment of judges. It also stripped them of their power to remove corrupt judges. When the new British-appointed Court of Common Pleas for the county of Worcester tried to sit in September, thousands turned out to block them. Of the estimated six thousand, about one thousand were armed. They stopped the court from coming to session and formed a "convention" that effectively took over, closing courts and freeing prisoners.

The weapons, which were not used, were unnecessary, since no armed force opposed them. Everywhere else in Massachusetts where the British tried to open a Court of Common Pleas, they were also stopped by huge crowds, which often had no weapons at all. The crowds were large enough to keep the courts closed, force the judges to resign, and keep the army helplessly at a distance.

The revolution had overthrown the government in Massachusetts without a shot being fired. Why, then, did the rebels turn to warfare? Sentiment was already strongly anti-British. John Adams wrote to Jefferson late in his life, "The revolution was in the minds of the people, and in the union of the colonies, both of which were accomplished before the hostilities commenced." So why was the war necessary? Jonathan Schell in *The Unconquerable World* astutely noted that participants in other revolutions had reached similar conclusions. The Romantic writer François René de Chateaubriand, who lived through the French Revolution, said almost the exact same thing: "The French Revolution was accomplished before it occurred." And Leon Trotsky, one of the authors of the Russian

Revolution, wrote, "The declaration of October 23 had meant the overthrow of the power before the government itself was overthrown."

So if revolutions are accomplished in the minds of the people, why must they be followed by force of arms? Why do almost all political theorists—not only Locke, Hobbes, and Rousseau, but later ones such as Marx and Lenin—insist that a revolution must be an armed movement? If the outbreak of war is inevitable, as seventeenth-century thinkers believed, history teaches the lesson that its inevitability does not rest, as they believed, on natural law, but on individuals incapable of conceiving of another path. Is the source of violence not human nature, as Hobbes contended, but a lack of imagination?

In the case of the American Revolution, could independence have been accomplished without warfare? The British gave up on America even though the Americans had scored very few military victories in the war, because they wanted to get on with other business, including their European wars, and could not afford to tie up military and money in these colonies any longer. But the path of disruption and protest had already been tying up British troops, costing Britain money, making the colonies unprofitable—the very reasons that Britain later gave up the war and negotiated peace. Colonies were supposed to earn, not cost. It seems quite possible that British withdrawal could have been achieved by continuing protest and economic sabotage.

———

The great lesson from the history of revolutions is that a shooting war is not necessary to overthrow the established power but is often deemed necessary to consolidate the revolution itself. By 1775 the revolutionaries may have, as Adams asserted, defeated the British, but they had not united the colonies. By the start of the shooting war, they had not even achieved the moment of enforced silence, because many, even supporters of independence, hoped that it was not too late to stop the war. There was in America, and especially in Philadelphia, which happened to be the seat of the Continental Congress, a vocal antiwar movement. French-born Anthony

Benezet, raised in a French Protestant family but a Quaker convert, stood on street corners passing out pamphlets against going to war and against slavery, both of which evils he insisted stemmed from the same impulse—a lust for wealth and power. He urged people to look within themselves for the causes of war and urged Patriots, as the revolutionaries liked to call themselves, to remember the meaning of Christianity and pray for reconciliation with the British.

In the winter of 1776 in Philadelphia, the majority of the Continental Congress was not prepared to call for independence. Some were still loyalists, and then there were the Quakers. The Quakers were deeply distrustful of people such as Adams, people who called themselves Patriots. They regarded these Patriots as extremely violent. By 1776 the Quakers had been confronting violent patriotism for some time. The nonpatriots noted that there was considerable criticism of British actions in the British Parliament and even expressions of sympathy for the American point of view. Was this not fertile ground for negotiations? The Patriots, including Quaker-born Tom Paine, were frustrated and angered by the Quaker position. John Adams had little patience for this point of view, believing negotiations would yield nothing and the only guarantee of American "liberty" was independence by force of arms. Franklin, a more experienced opponent of the Quakers, was even more impatient, regarding the entire debate a waste of time.

Leading the opposition to the restless Patriots was the Pennsylvania delegation, with its skilled parliamentarian John Dickenson. He was much respected as an early critic of British policy. An affluent London-trained lawyer with a Quaker mother, he lived in great luxury in Philadelphia, having married into an affluent Quaker family. He adopted both their wealth and some of their pacifism, though he held the rank of colonel in a Philadelphia battalion. Although anti-British and pro-independence, his repeated message to the Congress was that their goals could be achieved peacefully. Despite Lexington and Concord, the British retreat to Boston during which 200 British soldiers were killed, and their losing even more men in Boston at the battle erroneously labeled Bunker Hill, which actually took place on Breed's Hill, the British were reasonable men from similar traditions and surely negotiations would be possible, Dickenson insisted. Adams, however, argued for "powder and artillery." According to biographer David McCullough, Adams's irate assessment was that Dickenson must have been henpecked by his Quaker wife and mother.

It probably was too late. British troops had been killed and King George seemed determined to respond with his military might. Once this was clear, the Patriots got their moment of silence, an end to the debate about going to war. To oppose fighting was now equated with being a British loyalist. A loyalist in East Haddam, Connecticut, according to a loyalist report, was attacked by a mob for his politics. He was stripped of all his clothes, and pitch, so hot it burned his skin, was poured on him. He was then taken to a pigsty and had pig excrement rubbed on his body, thrown on his face, and forced down his throat. In Morristown, New Jersey, 105 men were sentenced to death for suspicion of being loyalists. Of them, 101 saved themselves by pledging loyalty to the revolutionary cause. The other four were hanged. Delaware sentenced a man accused of being a loyalist to be hanged and while still breathing cut down and drawn and quartered. Connecticut established the death penalty for loyalists, and in New York any statement deemed favorable to the loyalist cause was punishable by death. Even where it was not a capital crime, those who would not fight for the revolution were sometimes lynched, a practice named for Colonel Charles Lynch of Virginia, who would hang people by their limbs from a walnut tree in his yard until they screamed "Liberty forever!"

Families were bitterly divided. Benjamin Franklin turned against his own son, William, because he was a loyalist, and he did nothing to help him when William was thrown in an infamous underground dungeon in 1776.

After the outbreak of armed hostilities, the Pennsylvania Assembly fell largely into the hands of the revolutionaries, who required all white males of at least the age of eighteen to take an oath of "Renunciation and Allegiance." Such oaths were forbidden by most pacifist religions, but those who refused the oath could be greatly harassed. If they traveled, upon arriving in a new town they could be thrown in jail until they swore the oath. Those who did not take it were denied basic civil rights and sometimes imprisoned. Homes were confiscated from pacifists. More than most religious sects, the Quakers tried to enforce their beliefs, penalizing—even ejecting—members who in any way contributed to the war effort.

But not all pacifists were religious. Some secular Americans simply refused the draft and were imprisoned until they agreed to take up the cause.

The United States of America was founded by a war, and so it needed to be a "good war." The creation of this founding myth, the rewriting of history, began immediately after the war, while everyone with short-term memory knew otherwise. Collective amnesia was a small further sacrifice for nation building.

Most schoolchildren today are given the impression that the American Revolution was a relatively benign war. The worst thing that happened outside of the Continental Army being cold in the winter was the hanging of Nathan Hale, before which he got to make a speech asserting his willingness to be hanged.

In truth, the American Revolution was a brutal civil conflict filled with not only combat casualties but bitter feuds and abuses between civilians and between military and civilians. A higher percentage of the American population died in the Revolution than in any other war in U.S. history except the undeniably brutal Civil War. The Revolution, like most civil wars, was a war against civilians, a war in which women and children and the homes in which they lived were often deliberately and viciously targeted. Civilians would run in terror at the approach of either army. Homes were sacked and women were raped.

Saints were created and called "the Founding Fathers" but the Founding Fathers knew that such idolatry—Thomas Jefferson called it "sanctimonious reverence"—would harm the republic. They were far from the most progressive thinkers of the day. Slavery was their most celebrated flaw but they also set the stage for the genocide of some ten million American Indians, nor did they even entirely reject colonialism. They believed it was wrong to tax colonists who did not have representation in the legislature, but it was the tax, not the lack of representation, that was the grievance. Ironically, they repeatedly used words like *enslavement* and *slavery* to criticize taxation while at the same time accepting actual slavery. They were men of property, concerned with the issues of the affluent, including taxes. They wrote a Constitution that was far less progressive and enlightened than the laws adopted by the Pennsylvania colony half a century earlier. But they knew that they and their work were flawed. Jefferson, too, believed in the perfectability of humans, or at least that they would steadily grow wiser, and wrote that the Constitution should be rewritten in every generation to avoid having society "remain ever under the regimen of their barbarous ancestors."

We are indoctrinated by the image of "Patriots" who fought to free Americans, but some were brutal, some were bloodthirsty, and some agreed to fight only because of a lack of alternatives, which is one of the principal ways armies are raised. Most soldiers did not volunteer to fight for liberty. Immediately following the outbreak of war the rebelling colonies each issued their own draft laws, which included fines and sometimes imprisonment for conscientious objectors. But there also were many volunteers including, to John Adams's surprise, entire companies of Quakers.

Like all wars, the American Revolution filled its participants with horror. But its survivors were not allowed to commit such an unpatriotic act as discussing their experiences, and veterans were little cared for. Widows of enlisted men were not given pensions until 1832.

Typical of the postrevolutionary spirit was Charles Thomson, an Irish-born enthusiast of the American cause who was said to know as much about the war as any man alive. A friend of Benjamin Franklin and other key players, he served as secretary to the Continental Congress from 1774 until 1789. All during the Revolution he took detailed notes with the intention of writing a definitive history of the war. But after the war was over he decided not to write the book because too many good Americans would have their reputations sullied by accounts of their misconduct during the war. Before he died, at the age of ninety-five in 1824, he had all his notes burned.

Throughout the centuries pacifists have insisted that one of history's great lessons is that violence does not resolve disagreement. It always leads to more violence. Among those who reject nonviolence this contention is rigorously disputed. What cannot be disputed is that in the 1770s the American colonists chose violence over nonviolence, war over negotiation. A generation later, in 1812, the British and Americans went to war once more.

VII

PEACE AND
SLAVERY

The professed object of war generally is to preserve liberty and produce a lasting peace; but war never did and never will preserve liberty and produce a lasting peace, for it is a divine decree that all nations who take the sword shall perish with the sword. War is no more adapted to preserve liberty and produce a lasting peace than midnight darkness is to produce noonday light.

—DAVID LOW DODGE,
*War Inconsistent with the Religion of
Jesus Christ*, 1815

Even to those who reject the concept that there are just and un-just wars, it is clear that some wars have better arguments than others. The War of 1812 was not an easy one to sell. The Federalists, the more conservative of the two political parties in the United States, all opposed it. While it was supposed to be a war with the British, more than half of the enemy were Canadians, local militias raised in the loyal northern colonies. It was the first American war to spur a large antiwar movement, though like most effective antiwar movements this one drew as many objectors on economic grounds as it did on moral ones.

Those who objected on moral grounds were not all Quakers or from other peace sects. The man who is considered the first American peace activist was an affluent New York merchant and devout Presbyterian, David Low Dodge, who at the end of the war, in 1815, formed the New York Peace Society, the first nonsectarian peace organization in America history.

Born in Connecticut in 1774, he grew up during the violence and upheaval of the Revolution and later became a businessman in Hartford. Starting in dry goods, he became manager of Connecticut's first cotton mill in Norwich, and seemed to make money at everything he did. In 1802 he moved to New York, where he began his peace activities. In 1827, at the age of fifty-three, the self-made entrepreneur retired from business to devote his time to activism, until his death in 1852.

He mustered with a militia and, according to his autobiography, traveled with a pistol, prepared to defend himself during a rash of highway holdups. He claimed that he first questioned the consistency of considering himself a Christian and carrying a weapon when one night he nearly shot an innkeeper by mistake. He contemplated how he would have felt and how "God would have viewed the transaction" had he taken a human life. He spent several years studying the Christian position on the issue and eventually began to think the unthinkable: Was the American Revolution immoral?

In 1809 Dodge published a pamphlet that stated that even defensive war was not justifiable and that states that engaged in warfare were not Christian, were pagan and perhaps satanic. He cited the Gospels and the lives of early Christians to support his argument. The pamphlet sold out its thousand-copy press run in two weeks. It attracted both critics and admirers, and out of it came a circle of like-minded people who became the New York Peace Society, which was formed at the time of the outbreak of the War of 1812. But Dodge and his peace group, respecting the moment of silence in order to avoid the accusation of disloyalty, postponed official formation and publication of his tract against the war, *War Inconsistent with the Religion of Jesus Christ,* until 1815, when the war was over.

In the long history of such tracts against war Dodge's is one of the most thorough, cogent, and all encompassing. Like most of the American pacifists of the first half of the nineteenth century, he denounced war preparation, including maintaining an armed force, as inevitably leading to war. Dodge was influenced by the Revolutionary War experiences of his two half brothers who went off to fight at the ages of fourteen and sixteen, both of whom told of terrible battles, suffered sickness and privation, and died near the end of the war. He himself had childhood memories of the horrors of the Revolutionary War:

Who can describe the distress of a happy village suddenly encompassed by two contending armies—perhaps so early and suddenly that its inhabitants are aroused from their peaceful slumbers by the confused noise of the warriors more ferocious than the beasts that prowl the forest? Were it not for the tumult of the battle, shrieks of distress from innocent women and children might be heard from almost every abode.

He observed how war exploited the poor:

Very few, comparatively, who are instigators of war actually take the field of battle, and are seldom seen in the front of the fire. It is usually those who are rioting on the labors of the poor that fan up the flame of war. The great mass of soldiers are generally from the poor of a country. They must gird on the harness and for a few cents per day endure all the hardships of a camp and be led forward like sheep to the slaughter.

Dodge understood that most people were not used to the idea of questioning war and he saw this shake-up in the orderly nineteenth-century mind as his mission, pedaling his Society's pamphlets with the same salesmanship with which he had once pedaled dry goods. Being an adroit businessman, he presented economic arguments as well. "War is unwise," he wrote, "because it destroys property." Wars, he pointed out, are expensive and financed by taxes that affect first the merchant but are passed on to the consumer. "In times of war prices of the necessaries of life are generally very much increased, but the prices of labor of the poor do not usually rise."

It may be the poor who suffer most in war, as Dodge argued, but the New York Peace Society was a gathering of the propertied class, Dodge's circle—merchants, Wall Street brokers, philanthropists, and clergy. Early in its life, some Americans could see that the young United States, founded in war, as many states are, had a tendency toward perpetual warfare. By 1815 two generations had fought and it seemed possible that every generation of Americans would have its war.

Warfare produces peace activists, and they are likely to be found among the veterans of any war. Noah Worcester, a Unitarian minister who had fought as a volunteer in the American Revolution, became one of the first American veteran antiwar activists. Worcester could not find a publisher for his work, *A Solemn Review of the Custom of War*, probably because the country was at war at the time. Finally, as peace was being negotiated at the end of 1814, he self-published the pamphlet, which would become a classic of antiwar literature. The following year, he formed the Massachusetts Peace Society. War, Worcester contended, was "the grossest delusion that ever afflicted a guilty world."

With the war over, activists were free to launch a full-scale critique of warfare. In the early 1820s a Pennsylvania Peace Society was led not by a Quaker, but by Henry Holcombe, another veteran of the American Revolution. In 1828 an umbrella group was formed to unite all the peace societies, called the American Peace Society. Peace societies became an important intellectual force in nineteenth-century America, attracting speakers such as Ralph Waldo Emerson, who believed that humankind went to war because it was backward and undeveloped and that it would eventually abandon such practices.

As time distanced America from its founding experience, more and more Americans dared to speak out against the official version of the Revolutionary War. In 1839, Charles K. Whipple, a second-generation American born in 1808, wrote an antiwar tract titled *Evils of the Revolutionary War.* Some Americans were seeing fundamental problems with the new nation and suspecting the root of the troubles lay in how the nation was founded.

Two great moral arguments of the day merged, both concepts particularly centered in New England: nonviolence and the abolition of slavery. The leading voice of nonviolent abolitionism was William Lloyd Garrison.

Garrison, unlike earlier peace leaders, had emerged from poverty through self-education. By background he had little in common with anyone in Dodge's New York Peace Society or most of the other societies. Garrison's route out of poverty from the Massachusetts coastal town of Newburyport was journalism. The young man had already earned such a reputation for his articulation of the abolitionist cause that in 1829, Benjamin Lundy, the editor of a Maryland abolitionist newspaper, the *Genius of Universal Emancipation,* walked from Baltimore to Bennington, Vermont, to find Garrison and persuade him to become coeditor. Garrison's work in

Baltimore would earn him seven weeks in prison on a libel charge. In 1831 he began his own newspaper, *The Liberator*, which he continued publishing for the next thirty-five years. The paper never sold more than three thousand copies, but it was quoted and discussed everywhere. Its positions, Garrison's positions, were always clear and uncompromising and stated in bold language. He called for the immediate, unconditional liberation of all two and a half million slaves and called for the boycott of elections until slavery was abolished.

He rejected all violence, but he stirred people to action. Even if a curmudgeon by vocation, he was an enthusiast by nature. He loved taking long walks, and when he happened upon a place that he thought particularly beautiful, he would say, "Someone ought to build a hotel here!"

Perhaps because of his working-class background Garrison was something different from the pacifists before him. Though he had a deep knowledge of the Bible, he was a secular force, largely shunning established religions. He was not a pacifist. He was a nonviolent activist and used the force of his words and his charisma to call for people to act—to end slavery, to stop war, to let women vote, even to ban alcohol. But the abolition of slavery is the cause for which he is most remembered. He used to tuck his daughter into bed every night, reminding her that "the poor little slave child" did not have a warm comfortable bed such as hers.

Garrison, who once publicly burned a copy of the Constitution, called for the Northern states to secede from the Union because of the U.S. practice of slavery. He called the U.S. Constitution "a covenant with death and an agreement with hell." According to Garrison, the Constitution bound the North to the South in an "unholy alliance," and as long as Northerners remained part of the Union they could be drafted "at a moment's notice" to go south and put down a slave rebellion. The slave system in the South, Garrison stated, was guaranteed by "Northern bayonets." He was one of the first, and one of the few in history, to denounce the "Founding Fathers," who he said were led by their principles "to spill human blood like water, in order to be free." Garrison denounced the idea that good could come from the evil of killing. He was, in his day, one of the most admired and hated people in America.

In 1831, a small band of slaves in Virginia led by a fellow slave named Nat Turner attacked slave owners, killing at least fifty-five of them. This set off such hysteria—slave owners lived in terror that their huge black populations would rise up against them—that the state of Virginia actually proposed abolishing slavery, and the resolution nearly passed. Instead, the state chose a policy of repression. In the hysteria, almost two hundred black people who had nothing to do with the uprising were killed.

Even though it had nearly led to the abolition of slavery in the state, the violent uprising was completely unacceptable to many abolitionists, and especially to Garrison, who nevertheless was continually accused of having organized it. Threats were constantly made against his life and finally, in 1835, while attempting to deliver an antislavery speech in Boston, he was attacked by an angry proslavery mob. They chased him through the streets until he hid behind a pile of planks in the storage loft of a carpentry shop. About a dozen men found him there and were about to throw him from the second-story window when someone suggested his death be more prolonged. A rope was tied around him and he was lowered out the window down to a mob of several thousand, who dragged him by the rope through the streets while debating whether to hang him or first dye him with indelible black ink and tar and feather him. He was rescued at the last minute by the police. The affluent Wendell Phillips, watching from the windows of his law office, observed how this thin bespectacled young man with an already receding hairline never lost his composure throughout the entire ordeal. Garrison was prepared to be martyred for his beliefs. Phillips would become one of Garrison's closest collaborators.

Despite the frequent allegation that Garrison lacked organizing skills, strong and growing nonviolence and abolition movements developed around him in New England and throughout the country. By 1840, 200,000 Americans, including some Southerners, belonged to abolitionist societies. When Harriet Beecher Stowe's novel about a slave girl's escape, *Uncle Tom's Cabin,* was published in 1852, the 5,000-copy first edition sold out in forty-eight hours, and 300,000 copies were sold the first year alone.

Many of the most dedicated abolitionists identified Garrison as their inspiration, among them the daughter of a Medford, Massachusetts, baker, Lydia Maria Child. One of the first American women novelists, in 1824, at the age of twenty-two, she wrote the first historical novel published in the United States, *Hobomok: A Tale of Early Times.* Published two years before James Fenimore Cooper's better remembered *The Last of the Mohicans,* Child's book was unusual in its sympathetic portrayal of American Indians, as well as controversial in its exploration of interracial marriage. Though she published the book anonymously, her identity was soon discovered and the book made her famous.

By 1828, when she married David Child, she already had a blooming literary career. But her husband was a dreamy idealist, given to lost causes and the accumulation of debt. In order to raise money she wrote a guide to her own cost-cutting household methods, *The Frugal Housewife,* an enduring success and a regular source of income despite the fact that personally she loathed running a household.

When she was twenty-nine, Garrison began publishing *The Liberator.* Describing the effect Garrison had on her, she wrote: "Old dreams vanished, old associates departed, and all things became new." She became a militant leader in the nonviolent abolitionist movement, working closely with both Garrison and the Boston Female Anti-Slavery Society.

In 1833 she published *An Appeal in Favor of That Class of Americans Called Africans.* In the best traditions of nonviolence she did more than condemn slave owners; she wrote of the physical and moral harm that slavery caused both slaves and owners, and she placed blame equally on Northerners and Southerners. She also discussed that most taboo subject, interracial sex. In the introduction to the book, she noted: "I am fully aware of the unpopularity of the task I have undertaken, but though I *expect* ridicule and censure, I cannot *fear* them."

An Appeal destroyed her literary career. She was fired from her position with a literary magazine, her book sales dramatically declined, and her publisher dropped her. "O for a large heap of money to throw it into your laps!" wrote Garrison to David Child in 1836, bemoaning his inability to help them. She became editor of a New York–based weekly, the *National Anti-Slavery Standard*. Like Garrison and many other abolitionists, she believed their cause went hand in hand with that of the nonviolence movement. In an 1841 letter, she wrote: "It seems strange to me that any comprehensive mind can embrace one and not the other."

During the Civil War she helped get supplies to slaves fleeing the South and wrote a reading primer for them. After the war she published another novel, *A Romance of the Republic*, which advocated interracial marriage as a way of healing the race-torn society. She also took up the causes of woman suffrage and Indian rights, especially opposing the displacement of the Cherokee.

When Lydia Maria Child died in 1880, Wendell Phillips, who delivered her eulogy, described her as someone "ready to die for a principle and starve for an idea." Her literary career, destroyed by her political courage, was never resurrected, and today she is remembered for two things, *The Frugal Housewife*, and for penning the children's Thanksgiving jingle "Over the River and Through the Woods."

———

Another abolitionist inspired by Garrison was Elijah Lovejoy, a minister from Maine who moved to Illinois, where mobs destroyed his printing press four times to keep him from publishing an anti-slavery newspaper and finally killed him with five gunshots. Lincoln called the 1837 killing "the most important event that ever happened in the new world." But Garrison noted that Lovejoy had killed someone while trying to defend himself and was concerned about the direction abolitionism would take in the face of such violence.

Most abolitionists did not believe violence could be used in a good cause. From the seventeenth century to the mid-nineteenth century, there were few years in which at least one slave insurrection did not occur. Slaves tried anything they could conceive of to resist, including not only armed uprisings and plots to poison food supplies but mass suicide and traditional nonviolent approaches such as work stoppages and agricultural sabotage. The great majority of uprisings were violent and designed to spread fear. In most of history, people motivated by fear have not acted well. It is true that insurrections made the white establishment reconsider slavery. But, as in the Nat Turner incident, the slave owners generally rejected that solution and opted for a more ferocious brand of repression in order to instill fear in the slaves, which in turn also failed.

In Haiti a violent revolution overthrew slavery, defeated Napoleon, and established the first black republic. The war, which lasted from 1791 to 1803, was characterized by the most vicious, unrestrained, violent racial hatred. The French declared a "war of extermination" in which ships were employed as gas chambers for mass killing. The revolting slaves, for their part, tried to exterminate the whites. In the end Napoleon lost 50,000 troops—two entire armies. The new nation, product of the only successful slave rebellion, was so poisoned by hatred and fear that two centuries later it has still not recovered. It is also true that the fear engendered by the Haitian revolution caused many in Europe and the United States to rethink slavery. It caused the United States government, fearing the growth of its African population, to ban the import of new African slaves, but this led to increasing the slave population by such outrages as breeding farms. It is also true that the Haitian revolution, along with countless other uprisings, spurred the French and the British to end slavery, which is why it is not true, as is often claimed, that Europeans ended slavery without violence. But what finally put an end to British and French slavery was the development of the European sugar beet, an alternative to the Caribbean cane sugar for whose production that slave system had been created. This was why Lydia Maria Child's dreamy husband squandered their meager income trying to grow sugar beets in New England, but he failed to recognize the fact that cotton, not sugar, was the principal slave crop in the United States.

While slaves tried to terrify slave owners, the slave owners tried to make the slaves even more afraid than they themselves were. This led to an exchange of brutalities in which the slave owners were the inevitable victors, having used their imaginations to devise ever more hideous ways to kill and torture. Those lynched after the Nat Turner uprising were skinned. If slave uprisings had been nonviolent, would slave owners have acted better? It is impossible to say. All that can be said is, as in the case of the American Indians, it could not have been worse.

———

John Brown, born in 1800, was one of the rare white abolitionists who believed in the power of violence and the persuasive force of fear. Brown had operated a station on the Underground Railroad, violating U.S. law by helping escaped slaves flee to Canada. In 1855 he moved to Kansas to join five of his sons (from the twenty children he had fathered) hoping to win the emerging territory as an antislave state. In response to an attack by proslavers on the antislavery town of Lawrence, four of Brown's sons and two sons-in-law hacked five proslavery men to death with swords, slowly chopping them down while Brown watched. Brown, who was convinced that the slave insurrections were the key to ending slavery, said he wanted to "cause a restraining fear."

Brown never concealed his belief in violence. When he met former slave Frederick Douglass in 1847, he told the abolitionist leader that it was his intention to provoke and lead a war to end slavery. He told numerous abolition leaders about his plan to raid the army arsenal in Harpers Ferry, Virginia, and lead an insurrection of liberated slaves with the captured weapons. He tried to persuade Frederick Douglass to join him, but Douglass refused. Shields Green, a fugitive slave who had accompanied Douglass on his last meeting with Brown, did join in.

As the atmosphere of the nation became more and more poisoned with violence, it became harder to hold to a stance of nonviolence. Wendell Phillips and many of the other nonviolent abolitionists approved of violence in the defense of fugitive slaves. Some gave up on nonviolence after the fighting began in Kansas. In 1855 a letter was published in the *Anti-Slavery Standard* by Charles

Stearns, a conscientious objector who had gone to prison for refusing to serve in the Connecticut militia. Now he was in Kansas and, according to his letter, he had armed himself after ten days. His justification was the most common justification for war. He said that his adversaries were not human beings. They were, he said, tigers. "I always believed it was right to kill a tiger," he argued. Stearns pointed out that Jesus specified turning the other cheek "if a *man* smite thee." Stearns wrote: "When I live with men made in God's image, I will never shoot them; but these pro-slavery Missourians are demons from the bottomless pit, and may be shot with impunity."

"The Southern slave holder is a man," Garrison desperately insisted in the *Liberator,* where he attacked the sending of arms to Kansas, a step taken by congregationalist preacher and abolitionist Henry Ward Beecher, brother of Harriet Beecher Stowe.

Brown never discussed his plans with Garrison, the one abolitionist he was certain would not approve. They did meet in 1857. Garrison kept quoting the New Testament and Brown the Old Testament and they found little common ground.

Brown carried out the raid on Harpers Ferry in October 1859 with a force of twenty-one men, of which only five were black. He took the arsenal and in fact the town but then failed to take further action, and a detachment of marines led by Robert E. Lee arrived and overtook them the next morning. The first of Brown's men killed in the battle was Dangerfield Newby, a large and powerful ex-slave who had been freed by his white father. On Newby's body were found letters from his wife, still a slave some thirty miles from Harpers Ferry, who wrote that he had to free her soon because her owner was in financial difficulty and might sell her. "If I thought I should never see you, this earth would have no charms for me," she wrote and then pleaded, "Do all you can for me, which I have no doubt you will. I want to see you so much." After the raid Newby's wife was sold to an owner deep in the South. Shields Green, who had met Brown with Frederick Douglass, was captured later and hanged, at the age of twenty-three.

Brown was convicted of treason and hanged. After the initial cry of alarm, a surprising number of Northerners supported his actions, including some nonviolent abolitionists. Henry David Thoreau spoke of how Brown defended "the dignity of human nature" and Emerson compared him to Jesus Christ. Lydia Maria Child wrote that the Harpers Ferry incident "stirred me up to consecrate myself with renewed earnestness to the righteous cause for which he died so bravely." Even Garrison was supportive, though he took exception to the violence. In a speech in Boston on the evening of Brown's execution, Garrison said that although he had "labored unremittingly to effect the peaceful abolition of slavery . . . I cannot but wish success to all slave insurrections. . . . Rather than see men wearing their chains in a cowardly and servile spirit, I would, as an advocate of peace, much rather see them breaking the head of the tyrant with their chains."

C. K. Whipple defined the abolitionist stance in a pamphlet called *The Nonresistance Principle: With Particular Application to the Help of Slaves by Abolitionists.* This work praises Brown but departs from his approach by saying that while slavery must be resisted, no one has the right to take another life. Whipple called for a nonviolent slave rebellion, arguing that slaves had a moral obligation to "utterly refuse to any longer be a slave." He admonished: "Quiet, continuous submission to enslavement is complicity with the slave holder."

The nonviolent abolitionist movement faced the same problem that the Quakers had faced with regard to the Indians. While it is perfectly feasible to convince a people faced with the most brutal repression to rise up in a suicidal attack on their oppressor, it is almost impossible to convince them to meet deadly violence with nonviolent resistance. The crushing of a violent uprising is far easier to justify than the slaughter of unarmed people, and therefore nonviolence has more power. But most people, if they are going to die to fight oppression, feel better if they can take down a few oppressors first. Then if it all fails, they will at least have struck a blow. Enslaved African-Americans still felt this way after centuries of seeing violence fail.

Soon Brown's vision, a bloodbath without precedent, would be unleashed, and thousands of men would march south singing of

"John Brown's body." The problem was, they were not marching to free slaves.

———

The outbreak of a shooting war once again successfully silenced the nonviolence movement. They still raised their voices for abolition, but most no longer spoke out against war. Joshua Octogenarian, one of the few who never gave the war his support, attended a meeting of the American Peace Society in 1861 and complained that the group appeared to stand "for the vindication of war, rather than that of peace." Pacifists found all sorts of reasons to make an exception for this war: because it was really a police action against secession, because of the moral imperative, because the Southerner was not human. Lydia Maria Child wrote in an 1862 letter: "I abhor war and have the greatest dread of military supremacy; yet I have become so desperate with hope deferred, that a hurrah goes up from my heart, when the army rises to carry out God's laws."

Garrison, who had cared little about preserving the Union, became a Lincoln supporter. He said that although nonviolence would have been a better way to free slaves, once the shooting started, advocating nonviolence had become "impracticable." And this was true. History has shown, over and over again, that the activist who insists on nonviolence in times of war becomes hopelessly marginalized. Garrison had another cause, abolition, and he wanted to be heard.

While other abolitionists like Wendell Phillips were distrustful of the president's halfhearted approach to emancipation, Garrison, who had distrusted all government and urged people not to vote, suddenly was arguing for patience and understanding. He now characterized the war, Lincoln, and the Republican Party as "instruments in the hands of God." He predicted in 1861: "There will be desolation and death on a frightful scale, weeping and mourning, and lamentations for the slain and wounded in thousands of families—but if it shall end in the total abolition of slavery . . . it will bring with it inconceivable blessings."

Garrison, the Lincoln supporter, had joined the establishment after a lifetime of resisting its trappings. Was this what Thoreau meant when he wrote: "Wherever a man goes, men will pursue and paw him with their dirty institutions?"

———

The Civil War was begun with the standard arguments. In the South, slave-owning aristocrats tried to convince poor farmers who never had and never would own a slave, who were lucky if they owned a workable plot of land, that they were fighting to preserve "our way of life" against the Yankees, a label that implied less than human status, who were menacing barbarians invading sacred lands. Soon the Union Army indeed did invade with a barbarity that would do much to confirm this line of reasoning. Northerners were fighting Southerners, who likewise had already been established in many of their own circles as less than human. Though the Union Army spent most of the war invading the South, Northerners were convinced that the war was justifiable as a defensive war because the South had fired the first shots at Fort Sumter.

There were those for whom fighting to free the slaves became a holy war, including free blacks and many formerly nonviolent abolitionists. But there were many for whom fighting and dying for what they termed "a bunch of niggers" was unacceptable. Even many of Lincoln's loyal supporters were not interested in ending slavery and tended to have a higher regard for slaveholders than they did for the generally disliked abolitionists. George Templeton Strong, a wealthy New York corporate lawyer who became an ardent Lincoln backer, wrote in his diary in 1850:

> My creed on that question is: that slave holding is no sin. That the slaves of the Southern States are happier and better off than the niggers of the North, and are more kindly dealt with by their owners than servants are by Northern masters.
>
> That the reasoning, the tone of feeling, the first principles, the practices, and the designs of Northern Abolitionists are very particularly false, foolish, wicked, and unchristian.

In view of the strong antiabolitionist feelings throughout the country, the Democrats tried to tar the Republicans with abolitionism, and this made Lincoln and his party defensive on the slave question in ways that are no longer remembered today. The John Brown Harpers Ferry raid, coming just before the 1860 election year, was particularly awkward. Many tried to say Lincoln and the Republicans were behind the raid. George Templeton Strong began to rethink his Republican leanings. In September 1860, less than two months before the election, he noted in his diary: "I do not like the tone of the Republican papers and party in regard to the John Brown business last fall, and I do not think rail-splitting in early life a guarantee of fitness for the presidency."

During the campaign Lincoln was forced to repeatedly deny ties to John Brown and abolitionism and to assert that his position was not to question slavery in states where it already existed. He also maintained that there was no need to worry about a slave insurrection since the slaves lacked the prerequisite means of communicating between plantations and too many of them were happy with their kindly Southern masters. The happy slave was a persistent fantasy in the North that abolitionists such as Lydia Maria Child worked hard to dispel.

Lincoln knew that dying to free black people was not a sellable idea. The Union army would sometimes return escaped slaves to their masters, until the Republican-controlled Congress made it illegal to do so in March 1862. "My paramount object in this struggle is to save the Union," Lincoln said, "and is not either to save or destroy slavery. . . . If I could save the Union without freeing any slave, I would do it. If I could save it by freeing all the slaves, I would do it. And if I could save it by freeing some and leaving others alone, I would also do that."

Nor were most Southerners fighting for slavery, an institution with which few of them had any connection. They were fighting to drive out the invader. This worked for a while, though the South, too, had its war resisters and even its abolitionists. A small group of Southern Quakers in eastern Tennessee, who were both, spent a substantial part of the war in a large cave, its opening concealed by brush. War resisters in the Confederacy were sometimes subjected to beatings, repeated bayonet stabbing, being hung by the thumbs, and other brutalities for refusing to help fend off the savage invader. But by the beginning of 1865 the rate of desertion in the Confederacy was several hundred a day and Lee could barely keep an army together.

The first three years of the war, however, went very badly for the North, far worse than anyone had expected, and the casualties on both sides were horrific. By January 1863 about two hundred men a day were deserting the Army of the Potomac, and one in four soldiers was listed as absent without leave. Abolitionists constantly pressured the Republican Party on slavery. In the debut edition of *The Liberator,* in 1831, Garrison had attacked Congress for allowing slavery in the District of Columbia, writing, "That district is rotten with the plague, and stinks in the nostrils of the world . . . open to the daily inspection of foreign ambassadors." In the spring of 1862 slavery was abolished in Washington, D.C., so that at least the Union's capital was no longer a slave zone.

Lincoln, in need of a crusading cause, said that he would free the slaves "if I were not afraid that half the officers would fling down their arms and three other states would rise." The Union included border states, such as Maryland, where slavery was practiced, and Lincoln worried about holding on to them. He backed a plan to buy slaves for $400 each and send them to a repatriating colony, in either Africa or Haiti.

Lincoln was looking for a half measure that would address the great cause but not offend its opponents. For this to work he believed he needed to announce it on the crest of a Union victory. In desperation he considered Antietam to be that victory. On one day, September 17, 1862, considered the bloodiest day of the Civil War, 23,000 men were killed or wounded. The casualties were almost evenly divided, with slightly heavier losses on the Union side. But Lee's advance had been stopped, and so the Union claimed it as a victory. Five days later Lincoln issued the Emancipation Proclamation, a much misunderstood little document that reads: "All persons being held within any state, or any part of a state, the people whereof shall be in rebellion against the United States, shall be, thence forth, and forever free." Lincoln had only freed the slaves that he could *not* free, the slaves living in Confederate-held territory. Not a single slave was freed by the Emancipation Proclamation.

After a war has spent lives, people invariably demand that it be for something. It had to be for more than holding the Union together. Soon after Lincoln's inauguration to a second term the Thirteenth Amendment to the Constitution, outlawing slavery, was announced. After the war ended it was ratified by twenty-seven of the thirty-six states. Misssissippi was the only state to reject it and never reverse its decision.

Between 618,000 and 700,000 Americans, including 36,000 African-Americans, died in the Civil War, more than the total deaths from all other American wars combined. The $20 billion cost of the war was five times the total of government spending from its founding until 1861. One Confederate family, the Christians of Christianburg, Virginia, had eighteen family members killed. The Twenty-sixth North Carolina lost 714 of its 800 men at Gettysburg. The First Minnesota suffered an 82 percent casualty rate at Gettysburg. It was not unusual for a regiment to lose more than half its soldiers in a single battle—in a few cases in a matter of minutes.

The question is: Was this bloodbath necessary to free the slaves? Considering the fact that the war had not been fought for that purpose, it is hard not to wonder what could have been negotiated without warfare, if the will to end slavery had existed in the North. Most European countries had already negotiated the end of slavery and were sweetening with sugar beets and ignoring the anger of Caribbean planters. What is certain is that the great bloodbath in America freed the slaves in name only.

In May 1865, one month after the Civil War ended, the American Anti-Slavery Society met and lapsed pacifist William Lloyd Garrison faced a room of largely lapsed pacifists. Thinking, contrary to his lifelong beliefs, that violence had accomplished good work, he jubilantly announced that slavery had ended, their task was done, and he suggested that the organization disband. But Frederick Douglass insisted that the Southern black could not vote and was not free. The majority, many of whom, not coincidentally, later moved on to the cause of women's right to vote, agreed with Douglass that without full rights of citizenship, blacks were still not free.

Because war had been the tool of emancipation, emancipation became the task of a hated occupying army. The Union army was the only guarantor of black rights and safety in the South. There is some question of how much enthusiasm the army brought to this task. Though most soldiers had never been interested in the cause of abolition, freeing slaves had been done enthusiastically because it was seen as a punitive measure against the South. In fact, one of Lincoln's arguments in favor of emancipation had been that it would hurt the Southern economy. But continued strife to gain the black man his rights was of little interest. Lincoln was dead and the new president, Andrew Johnson, was from a slave state, Tennessee. Johnson appointed proslavery governors to rebuild the conquered states, and these men formed governments that despite the Thirteenth Amendment were proslavery. The new legislatures of the Southern states passed laws to deprive African-Americans of rights and economic opportunities including "black codes," laws under which a former slave could be arrested if unemployed and ordered to pay large fines. Former slaves who could not pay the fine could be hired out for labor until the fine was earned, making them, in effect, slaves again. Black children were forced to work as unpaid or barely paid apprentices. Blacks were barred from owning land.

By 1870, military rule ended and the Confederate states were back in the Union. Republican majorities in these states supported by black votes evaporated as blacks were deprived of their voting rights and the states fell under solid Democratic control. State governments in the one-party system were free to pass laws inhibiting black rights, and where they couldn't because it contradicted federal law, they could use force. Violent groups such as the Ku Klux Klan made it extremely dangerous if not impossible for blacks to vote or attend schools.

In reality, the only right Southern blacks had gained was to go north, if they could get together the money, live in an urban slum, and labor long hours at an underpaid, dangerous job created by the new industries that the Civil War had developed. With 700,000 dead and hundreds of thousands more maimed, the agenda of "freeing the slaves," emancipating the African-American South, was deferred, left on a back burner for future generations in another century. One hundred years after the Emancipation Proclamation, Martin Luther King Jr. stood at the monument to Abraham Lincoln in Washington, D.C., before an unprecedented rally of hundreds of thousands and promised the black people of America that someday in the future they would see the fulfillment of the words of the old spiritual, "Free at last! Free at last! Thank God almighty, we are free at last!"

Someday, but a century after the great bloodletting it still had not happened.

VIII

THE CURSE
OF NATIONS

Our country is the world, our countrymen are all mankind. We love the land of our nativity only as we love all other lands. The interests, rights, liberties of American citizens are no more dear to us than are those of the whole human race. Hence, we can allow no appeal to patriotism, to revenge any national insult or injury.

—WILLIAM LLOYD GARRISON,
Declaration of Sentiments adopted by
the Peace Convention of Boston, 1838

For the Paris World's Fair of 1867 a group of notable writers including Alexandre Dumas, George Sand, and Victor Hugo put together a book for which Victor Hugo wrote the introduction. They sent a copy to the National Library in Paris, which they said would one day be called the Library of the United States of Europe. Hugo and the others, like numerous European intellectuals of the time, had looked at their recent history of constant warfare and concluded the problem lay in the existence of the nation-state. They reasoned that if there were no countries there would be no wars. They imagined a future in which Europe would be at peace because it would no longer have nations.

The Hugo piece begins:

> In the twentieth century there will be an extraordinary country. This will be a large country but that will not keep it from being free. It will be illustrious, wealthy, thoughtful, pacifist, cordial to the rest of humanity. It will have the gentle gravity of an elder. It will be bemused by the glory of conical projectiles and will find it hard to understand the difference between an army general and a butcher; the royal purple of one not seeming very different from the red of the other. A battle between Italians and Germans, between the English and the Russians, between the Prussians and the French—this will seem to it as a fight between Picts and Bourgognes might seem to us. It will consider the wasting of human blood pointless.

But the old ways of thinking would not vanish suddenly. In his piece on this new future, Hugo asserts without hesitation that the capital of this new superstate would be Paris.

The 1867 World's Fair was an unlikely spot to announce the end of nationalism. It was the first world's fair to be financed by a government—the French government—rather than by private initiatives. Napoleon III, whom Hugo bitterly opposed, intended it as a showcase for his Second Empire, especially Haussmann's reconstruction of Paris. Émile Zola, another opponent of the French government, thought the fair was being used to repress the opposition. But all the other nations of Europe and the world also used it to show off. Heads of state arrived with great pomp. Aristocrats could enter free of charge through a special gate. The U.S. government sent its first official world's fair delegation.

The Germans used their exhibition hall to show off innovations in artillery. A new cannon by Krupp, at the time the largest artillery piece ever built, turned out to be the most popular exhibit at the fair. The French were not troubled by the German choice for their exhibition space. The French press concluded that the fact that the Kaiser himself came to show off the cannon was proof that warfare between France and Germany was a thing of the past.

Three years later the Franco-Prussian War began, a French-German conflict that led directly to World War I, which led directly to World War II. Wars never end warfare, they lay the groundwork for the next.

———

The idea of a united Europe was an old one. Dante, Erasmus, Rousseau, and Immanuel Kant are among the many European intellectuals who called for a pan-European organization to resolve problems and avert war on the Continent. A common currency, eliminations of customs and tariffs, a unified educational system, a reduced emphasis on war in the teaching of history were all common proposals. Late-nineteenth-century idealists imagined resolving the troubling issues on the French-German border with an international tribunal in The Hague. There was a growing cry for an end to secret diplomacy and a demand for treaties to be openly debated and ratified by legislatures. Peace activists talked of uniting Europe with capitals in Brussels and Strasbourg. Strasbourg was meaningful because it was the capital of Alsace, a disputed territory at the heart of the Franco-German conflict.

Until the nineteenth century such utopianism was the stuff of dreamers. But in 1815, with both the Napoleonic Wars and the War of 1812 having ended, Europeans and Americans were war weary and looking for new answers. As in the United States, peace societies with sizable memberships started forming in Europe. The first pan-European peace organization was established in Geneva in 1830, but it lasted only nine years, until the death of its founder, a Protestant aristocrat, Jean-Jacques de Sellon. Sellon had been an outspoken opponent of capital punishment, which gave him credentials as a progressive—credentials he was much in need of considering his ambivalence toward the French Revolution and participation in the Napoleonic state. In 1829 he began thinking that the cause of peace was closely related to that of capital punishment and he became a peace activist.

The Paris Peace Conference of 1849 was attended by a diverse group that included several captains of industry, the Archbishop of Paris and the Grand Rabbi of France, and Alexis de Tocqueville. The American delegation got a great deal of press attention thanks to the attendance of William Wells Brown, an escaped slave, Underground Railroad activist, and close collaborator of William Lloyd Garrison. Victor Hugo, a last-minute replacement for the ailing archbishop, presided over the events, which he opened with his proposal for a United States of Europe.

But neither complete disarmament nor even arms reduction was popular at the conference. The archbishop asserted the threat of barbarians in the north, i.e., Russia, while others compared the Austrian army to fifth-century Goths. Surely, it was argued, such dangers required armed resistance.

———

Another pan-European peace organization, the International League of Peace and Freedom, was formed in 1867 in Geneva. Its founding meeting was scheduled to follow directly the meeting of the Socialists' First International in Lausanne, the assumption being that the delegates to the International could save on travel expenses by staying over for the peace convention. But the General Council of the First International, at the request of Karl Marx, asked its delegates to oppose all proposals at the International that were in favor of the Peace League and to decline any official participation in the peace conference.

A quotation frequently cited from Marx's seminal work *Capital* is "Violence is the midwife of every old society pregnant with a new one." *Violence* is sometimes translated as *force*, which is not exactly the same thing. In its context, a chapter titled "Genesis of the Industrial Capitalist," it is not clear if he is speaking of revolutionary force or violence. The statement is followed by a discussion of the brutality of colonialists, and Marx seems to have been saying that the move for social change is spurred by the force inherent in the established capitalist system. Marx believed that although violence was inevitable, by his day already an old Hobbesian notion, that it was also irrelevant to the outcome. The outcome depended on socioeconomic conditions, just as the power of the state, according to him, did not reside in its ability to use force but in its control of the means of production. So it appears Marx believed that the violence, or force, that was a midwife came from the old order and was not necessarily the prescribed path for a new one.

But Marx, like most people interested in power politics, was deeply distrustful of pacifists. Ironically, one of his concerns was that pacifist policies might leave Western Europe vulnerable to an invasion by Russia.

Nevertheless, among the six thousand people who attended the Geneva peace conference were numerous delegates from the just concluded International, among them the Russian anarchist Mikhail Bakunin. Bakunin, despite his interest in the peace conference, believed in the violent overthrow of states as a means of establishing true liberty. Another participant in both the International and the peace conference was William Randal Cremer, a lifelong champion of the idea of international arbitration, who would become the first blue-collar worker to win a seat in the British House of Commons. In 1887 he persuaded 234 members of Parliament to sign a resolution addressed to the president of the United States asking the two countries to sign a treaty pledging that any future disputes between London and Washington would be turned over to international arbitration. In 1903 he won the third Nobel Peace Prize.

One of the starring attractions of the conference was its official host, Giuseppe Garibaldi, who was not nonviolent but was enormously popular. He had fought in Italy, in Latin America, against the French, later for the French, and had conquered Sicily and Naples and unified much of Italy. In his opening statement he attacked the pope, and Catholics, as a result, were so angered that they threatened to storm the conference. On the second day, Garibaldi called for the violent overthrow of the Italian Papal States. By the third day, Garibaldi had vanished. It was later understood that he had returned to Italy to overthrow the Papal States.

Amand Goegg, a German pacifist, warned that a Prussian-led united Germany would be a threat to world peace. The Socialists insisted that there could be no peace until a classless society was constructed. Nevertheless, the conference called for the abolition of standing armies. It also called for an end to racism, for the right to work, for the establishment of a United States of Europe, and for an international organization to work toward these goals. It would remain a vigorous organization for the rest of the century.

With the outbreak of the Franco-Prussian War in 1870, as with all wars, dialogue ended. Once Paris was surrounded by the Germans, few champions of peace spoke out. One French writer, Edmond Potonié, organized a small peace movement, which covered the besieged city with posters calling for talks with Germany. In Britain, William Randal Cremer campaigned to keep the British out of the war.

After the war, peace activities resumed with an even greater sense of urgency. A network of national peace organizations brought together several thousand activists from Western Europe, the United States, Argentina, Japan, and Australia. Though Leo Tolstoy, who had become a renowned peace activist with a strongly religious perspective, belittled such meetings, saying that men simply had to refuse to fight and that international conferences meant nothing. There were twenty such conferences between 1889 and 1914. In 1896 supporters of an Alsatian Jewish captain in the French army, Alfred Dreyfus, who had been convicted of treason, discovered proof of his deliberate framing by right-wing anti-Semitic elements in the French military. The Dreyfus case pitted the left against a militarized right and did much to drive the left-wing establishment toward antimilitarism, not only in France but also in neighboring countries. Some even dared hope that war might become a thing of the past among civilized nations. "Civilization is peace, barbarism is war," wrote Frédéric Passy, the French economist and peace activist who in 1901 received the first Nobel Peace Prize.

In 1889, Baroness Bertha von Suttner, an Austrian peace activist and the author of numerous books of fiction and nonfiction, published an antiwar novel. The novel, titled in the original German *Die Waffen nieder, Lay Down Your Arms,* is a semiautobiographical story of an aristocratic Austrian woman from a military family who, like the author, comes to realize that the values with which she was raised were grievously mistaken. The wars written about were ones that von Suttner had lived through and carefully researched, and so the novel had the ring of truth for contemporary readers.

Die Waffen nieder, translated into many languages, was a huge international success—it was to the peace movement what *Uncle Tom's Cabin* had been to abolitionists. A film was even made of it in 1915. A former secretary to Alfred Nobel, von Suttner frequently wrote him about peace issues and is thought to have been a key influence in his decision to include the peace category among the Nobel prizes—a distinction she was awarded in 1905.

Surely peace was imminent. Andrew Carnegie, the Scottish-American industrialist, set up a $10 million fund to abolish war. He was so confident of achieving his goal that he began discussing what other "degrading evil" should be taken on after war was banished.

———

But European peace advocates began to be disillusioned with what had been to many the new bright hope, the United States. In 1897 the U.S. Senate rejected William Randal Cremer's Anglo-American treaty for arbitration. Worse, the following year the United States went to war with Spain, a war in which the U.S. interest was clearly the acquisition of colonies. To progressive Europeans it was a shocking blow to see the United States engaging in European-style colonialism. Among the European left, the U.S. image has never recovered. In the French Chamber of Deputies, the Socialist deputy, Francis de Pressensé, said with sadness, "The seductions of imperialism are drawing the United States toward the abyss where all the great democracies have found their end."

The United States also had its own domestic critics, notably the author Mark Twain, who, as a leading figure in the Anti-Imperialist League—formed to oppose the Spanish-American War and especially the invasion of the Philippines, which was actually carried out after an armistice with Spain was signed—wrote numerous passionate attacks. In his "The War Prayer," he argues that praying for victory is praying for killing. Despite his fame, it was rejected for publication. One of his published comments on the Philippines invasion was: "We have pacified some thousands of the islanders and buried them. And so by these Providences of God—and the phrase is the government's not mine—we are a World Power." Indeed it is hard not to notice how inauspiciously the United States assumed world leadership. In the lengthy annals of fatuous war justifications, U.S. president William McKinley deserves special mention for his 1903 explanation to a group of fellow Methodists of why he had sent 70,000 troops to crush local opposition to the U.S. occupation of the Philippines:

When next I realized that the Philippines had dropped into our laps I confess I did not know what to do with them. I sought counsel from all sides, Democrats as well as Republicans, but got little help. I thought first that we should take only Manila; then Luzon; then other islands, perhaps also. I walked the floor until midnight; and I am not ashamed to tell you, gentlemen, that I went down on my knees and prayed Almighty God for light and guidance more than one night. And one night late it came to me this way—I don't know how, but it came . . . that there was nothing left for us to do but to take them all, and to educate the Filipinos, and uplift and civilize and Christianize them, and by God's grace do the very best we could for them, as our fellow men for whom Christ also died.

There at the dawn of the twentieth century, "the American century," was Augustine's just war with the spin of Urban II and a little colonialist Hobbes and Locke mixed in along with that peculiar American twist that was now to be heard over and over again, that we will do them the kindness of bringing our ways to them and thus making their lives better.

In 1904 the French and British governments put aside centuries of animosity and signed a treaty agreeing to have their future treaties and disputes arbitrated in the Hague. But historians argue over whether this was a genuine step toward world peace or simply a realignment of alliances. Was France merely correcting the British neutrality of the Franco-Prussian War? In the future, the two countries would go to war not against each other but in concert against Germany.

———

In August 1914 the cataclysm happened, and it happened for all the reasons the peace movement had warned against—a buildup of armaments, secret negotiations, and excessive flag-waving nationalism. Europeans were so used to the idea of war that at first they did not understand that this one would be different. The difference lay in all the armaments that had been developed in the industrial age—huge destructive artillery pieces, chemical weapons, rapid-fire machine guns, and airplanes, which Bertha von Suttner, who died two months before the outbreak of war, had campaigned to exclude from being used as weapons.

Prophetically, von Suttner had described the atmosphere in the early weeks of the war in *Die Waffen nieder*:

> Nothing else was spoken of in rooms or streets, nothing else read in the newspapers, nothing else prayed about in the churches. Wherever one went one found everywhere the same excited faces, the same eager talk about the possibilities of war. Everything else which engaged the people's interest at other times—the theatre, business, art—was now looked on as perfectly insignificant. It seemed to one as if it were not right to think of anything else whilst the opening scene in this great drama of the destiny of the world was being played out.

And in this atmosphere the peace movement, along with the peace, all but vanished. Philosopher Bertrand Russell had collected the signatures of more than sixty Cambridge fellows and professors on a petition demanding British neutrality in the event of war. But once war was declared, most of them withdrew their opposition. Russell later wrote of the war: "The prospect filled me with horror, but what filled me with even more horror was the fact that the anticipation of carnage was delightful to something like ninety percent of the population." Russell's figure was probably exaggerated since it did not take into account the large percentage of the population silenced by the fear of being ostracized, or accused of cowardice or a lack of patriotism.

But as the war plodded on, bloodier and more interminable than anyone had imagined, more and more people spoke out. A coalition of the intelligentsia and the left-leaning labor movement, the same coalition the peace movement had tried to put together in Geneva in 1867, formed. This was especially true in Britain, where the labor movement had the slogan "A bayonet is a tool with a worker on each end."

In October 1915 the British attempt to break through the German lines at Loos had collapsed after costing 60,000 lives. In January the true nature of the war became clear when the remainder of British, French, Australian, and New Zealand troops sent to attack Turkey were evacuated from the Turkish peninsula of Gallipoli with 250,000 casualties. No weapon or invasion was going to provide an alternative to trench warfare in Europe. For all the new technology, this tactic depended on manpower. After the second battle of Ypres in May 1915, when the Germans first used gas and killed 30,000 British soldiers, the height requirement for British volunteers was lowered from five foot five to five foot three. The British had more volunteers than uniforms and they came in faster than they could be trained. But it was obvious that volunteers alone would not be enough.

For centuries Britain had been engaging in nearly constant warfare without a popular outcry loud enough to stop it because ever since 1660, after a civil war and several decades of protest against the royal prerogatives of warmaking, the British military had become completely voluntary. The British had learned from their seventeenth-century experience a lesson the Americans did not fully embrace until after the Vietnam War—that wars do not have to be sold to the general public if they can be carried out by an all-volunteer professional military. But in 1916, for the first time since 1660, the British government proposed a draft, and despite its unpopularity, it was passed by Parliament, after much debate and numerous revisions, with only thirty-five dissenting votes.

The Independent Labour Party had been against the war from its outbreak. But once the draft was established, most other labor organizations also voiced their opposition, at least to the draft if not the war. So did many other people. Vanessa Bell, the painter sister of Virginia Woolf, said that the absence of a draft was the one advantage of being British. "But if that goes I don't see any reason for bringing up one's children to be English." With the draft a whole subclass emerged, often treated more as a subspecies, called the "conchies"—the conscientious objectors.

The dissident voice of the conchies was offensive to the British propaganda machine of war, and so the attention that these 16,500 men received belied the sad reality that they represented less than half of 1 percent of draftees. Those who refused to serve were given five-minute hearings after which they were either granted conscientious-objector status, were sentenced to hard labor in prision, or agreed to go into the military. Those who announced a crisis of conscience after already being shipped to France received worse treatment, such as being chained to a post or artillery wheel for two hours and being put to hard labor in chains. The military authorities felt themselves to be very progressive for enacting such punishments as a replacement for the traditional flogging. Some conscientious objectors were sent to France to be abused.

Many writers volunteered to serve and then wrote of the horrors they witnessed, creating an enduring antiwar literature. Some, such as poets Rupert Brooke and Wilfred Owen, did not survive the war. Owen died in combat in the last week of the war. Siegfried Sassoon, who wrote some of the angriest poems, with near-lyrical passages about uncaring generals and slaughtered youth, was a decorated officer who tossed his Military Cross into the Mersey while on leave, an act that led the War Office to conclude he was "a lunatic."

The celebrated Bloomsbury Group was famously antiwar, though their stands ranged from that of economist John Maynard Keynes, who supported the government enough to work for it, to that of art critic Clive Bell, husband of Vanessa, who was its most outspoken opponent. He wrote a pamphlet, *Peace at Once,* which he later boasted was gathered and burned in the streets of London by the Lord Mayor. This was only a slight exaggeration. The pamphlet was published by the National Labour Press in Manchester, which was later raided by the Manchester police, leading to a famous freedom-of-the-press case. During the course of the trial the local magistrate had all confiscated material burned, including 1,642 copies of Clive Bell's antiwar pamphlet.

Most wars have shaky rationales, but the justifications for World War I were particularly thin, and Bell disposed of them in fifty-six pages. He argued that the survival of Britain was not at risk, that honor was an absurd reason for killing, that wars do not promote peace or democracy, and, prophetically, that "smashing the Germans" was not a good idea for the future of Europe nor anything that any sane person would want to die to accomplish. He reduced World War I to a squabble between ruling classes:

> That the ruling class in Germany would like to smash the ruling class in England, I do not doubt. It is a peculiarity of ruling classes that they want always to be smashing each other. It should be the task of democracies to see to it that they smash nothing more precious.

Standing against the war had a price. In 1917, the British government denied Bertrand Russell a passport in order to prevent him from giving a lecture series at Harvard, and at the same time he was removed from his lectureship at Cambridge. In 1918 he was sentenced to six months in prison for statements opposing the presence of a U.S. military base in England. While serving his sentence he worked on his later celebrated books, *The Analysis of Mind* and *Introduction to Mathematical Philosophy*, and read some forty books, including Lytton Strachey's now classic biography, *The Eminent Victorians*. Strachey himself had lost his position on *The Spectator* for his antiwar views.

In the United States the antiwar movement flourished until 1917, when the Americans entered the war. Suddenly laws were passed equating the expression of antiwar sentiments with espionage. Those who denounced the war could be sentenced to as much as twenty-five years in prison, yet 142 were sentenced for life and 17 were sentenced to death, though the executions were never carried out. Many thousands were so badly beaten and abused in prison in attempts to force them to change their stance, that at the end of the war only 4,000, about a third of the men who had said they would not serve, remained hard-and-fast conscientious objectors. The government allowed gangs to beat and even tar and feather war resisters and force them to kiss the flag. The American press, like that in Britain, belittled war resisters. Former president Theodore Roosevelt, speaking at the Harvard Club, called them "sexless creatures." Antiwar movies and books were banned, while people flocked to prowar propaganda designed to instill hatred of Germans.

———

Even in the early years of World War I stories of low morale were abundant. Corder Catchpool, a British conscientious objector who worked with an ambulance crew, wrote in a letter from France in November 1914 that he saw hope for the world in the attitude of the soldiers. He would talk to them about how the ruling class made the war and they suffered for it. "They almost always see the point," he noted. When the draft was instituted, Catchpool returned to England specifically in order to refuse it and go to prison. In 1917 he was being guarded by combat veterans with whom he debated. He wrote: "Not one has any delusions left about the war, such as one meets everywhere from civilians at home, and every man of them wants it to end, and doesn't care a toss how it is arrived at." By 1917, desertion was widespread in the armed forces of all of the fighting countries.

Finally, an exhausted and bankrupt Germany came to terms while its army still held parts of France, and the war ended without anyone having won a strategically decisive battle. George Bernard Shaw had visited the western front, where he reported that being a soldier was "soulless labor." He wrote of the average soldier: "He only hands a shell or pulls a string. And a Beethoven or a baby dies six miles off." According to conservative estimates, ten million had been killed and another twenty million wounded.

IX

A FAVORITE
JUST WAR

The kind of pacifism that does not actively combat the war preparations of the governments is powerless and will always stay powerless. Would that the conscience and common sense of the people awaken!

—ALBERT EINSTEIN,
speech in New York,
December 14, 1930

World War I had given war such a bad reputation that, for a moment, most people turned against it. Both youth and veterans groups had large numbers campaigning for peace. Quakers and other antiwar religious groups were expanding in Britain and the United States. The works of European writers and artists, as well as Americans such as Ernest Hemingway and John Dos Passos, had strong antiwar overtones as did many of the painters such as the German war veteran Otto Dix. The most startling contribution, perhaps the best antiwar novel ever written, was Erich Maria Remarque's *All Quiet on the Western Front*, published in 1929. That year it sold three and a half million copies in German and the twenty-five languages into which it was translated. "It should be distributed by the millions and read in every school," said the French newspaper *Le Monde*.

Remarque was a veteran of the war, and his novel was the simple story of a group of German schoolboys whose teacher instructed them on the value of patriotism and urged them to enlist; after arriving at the front, they endure all the horrors of the war and are killed, one by one.

> There was indeed one of us who hesitated and did not want to fall into line. That was Josef Behm, a plump, homely fellow. But he did allow himself to be persuaded, otherwise he would have been ostracized. And perhaps most of us thought as he did, but no one could very well stand out, because at that time even one's parents were ready with the word "coward": no one had the vaguest idea what we were in for.

For once veterans were telling the truth about the war they fought in and hated. And now the public was not only listening, they were applauding. At the same time more academic writers were publishing exposés—for example, Arthur Ponsonby's *Falsehood in Wartime* and Sir Philip Gibbs's *Now It Can Be Told*—on how the leaders on both sides had lied to the public. Even Woodrow Wilson, the president who brought the United States into the war, later said, "Is there any man, woman or child in America . . . who does not know that this was an industrial and commercial war?"

The Permanent Court of International Justice finally opened in The Hague in 1922. The long-awaited international organization, the League of Nations, was established after the war and successfully averted war between Sweden and Finland in 1921 and between Bulgaria and Greece in 1925.

How could populations ever again be duped into such slaughter? How could young men ever again be cowered into killing the way the German boys in Remarque's book were? But four years after *All Quiet on the Western Front* was published, Hitler came to power and ordered the novel burned while its author fled to Switzerland.

Still, antiwar sentiment was running high, especially in the United States. The peace movement was becoming mainstream. Leading scientists such as Albert Einstein were outspoken pacifists. Christian clergy were coming forward to vow that they would never again commit the sin of backing war. In 1935 the Central Conference of American Rabbis mailed a questionnaire asking its membership of Reform rabbis if they would in the future refuse to support any war. Ninety-one said they would, thirty-two agreed with certain qualifications, and only thirty-two said no. Women obtained the right to vote in the United States in 1929, one year after women were granted full rights in Britain. Where women had the vote, especially in the United States, that vote was considered, as it still is to some degree, an antiwar vote. The Women's International League for Peace and Freedom, a small group founded in 1915, grew into a major peace organization with 120 branches around the country. College students became vociferously antiwar, organizing large demonstrations.

When, following an unsuccessful coup d'état attempt in 1936, Fascists attempted to take over Spain by force, with military assistance from Hitler and Mussolini, neither the British nor the Americans, nor even the Socialist government of France, was prepared to go to war to save the Spanish Republic. But many idealistic volunteers, even some antiwar activists, went. Among them was David Dellinger, a student from an affluent old-line Boston family who had been impressed by Socialist leader Eugene V. Debs's stand against World War I, for which Debs had gone to prison. Dellinger graduated from Yale in 1936 and won a graduate fellowship at Oxford. Before going to Oxford he toured Fascist Europe, traveling to Spain, Italy, and Germany. In Spain he was very moved by what he saw as "the people's struggle" against fascism and even considered fighting for the Republic. But he saw the Communists shooting at the Trotskyites and the anarchists shooting at everybody, including at him as he rode in a car in Barcelona. "I knew," he wrote, "that I had to find a better way of fighting, a nonviolent way."

It was on the German leg of his trip that his convictions solidified. What he found in Germany was American corporations. "General Motors, ITT, and Ford come most readily to mind as having plants protected by Hitler, but there were others." Nazis would cite these investments when arguing to Dellinger that he should support them. Anti-Nazis would cite the plants and U.S. support when complaining of their difficulties in opposing the Nazis. He would go to bookstores and ask for works by the nineteenth-century lyric poet Heinrich Heine. He knew that Heine had been banned because he was Jewish, and Dellinger, fond of bookstores and their owners, thought this would begin a dialogue. Sometimes it did; many were anti-Nazi and complained about U.S support for the Nazis. Dellinger remembered the words German writer Thomas Mann had written in his diary two years earlier, in 1934:

> Russian socialism has a powerful opponent in the West. Hitler, and this is more important to Britain's ruling class than the moral . . . climate of the continent. . . . While horror of Hitler's methods is great . . . the governor of the Bank of England was sent to the United States to obtain credits for raw materials for Germany, i.e. armaments credits.

To the captains of banking and industry in the United States, Britain, and France, and to the political leaders who supported them, the enemy was not fascism, it was communism. When Dellinger got to Oxford, he met a German Rhodes scholar, a dedicated anti-Nazi who somehow had gotten through Nazi screening. Dellinger was surprised to discover that this German student was far more anti-Nazi than many of the upper-class teachers and students at Oxford. Through this student he met an entire network of anti-Nazi Germans who complained bitterly that no one would help them and that, in fact, the Americans and the British were helping the Nazis. These opponents of Nazism were strongly opposed to another international war and wanted to see the Nazis overthrown from within by a German movement. Dellinger made his last trip to Germany with his Boston Brahmin parents. At the border, a uniformed guard gave the Nazi salute as he stuck a swastika on their car. In plain view of the guards, Dellinger angrily removed the sticker. Though the guards said nothing, his parents were upset. They pointed out that none of the other American tourists objected to swastikas on their cars.

In 1937, Neville Chamberlain had been installed as prime minister at the head of a Conservative British government. He believed, and few today would argue with this, that Germany had been ill treated by the punitive terms imposed upon it at the end of World War I. Chamberlain thought this could be the basis of negotiations. But the British had other common ground with the Nazis. Chamberlain's foreign secretary, Lord Halifax, wrote in his diary that he had told Hitler, "Although there was much in the Nazi system that profoundly offended British opinion, I was not blind to what he [Hitler] had done for Germany, and to the achievement from his point of view of keeping Communism out of his country." Hitler had accomplished this by murdering or placing in concentration camps every leftist he could find.

In 1938 Hitler violated the World War I peace agreement by annexing Austria. In September the Chamberlain government, along with the French government of Édouard Daladier and Mussolini's Italian government, signed the Munich Agreement, which gave the ethnically German region of Czechoslovakia, the Sudetenland, to Germany in exchange for an agreement that this would end Germany's plan for German ethnic unification. The Czech government had not even been invited to the conference.

The Munich pact proved an infamous failure, and "appeasement," the name given to the Chamberlain policy, has become a curse word. The Western policy of constant aggressive hostility toward the Soviet Union—the postwar Cold War—was fueled by leaders who wanted to be certain that they did not commit the sin of appeasement. Robert McNamara, who as Secretary of Defense for Presidents Kennedy and Johnson was one of the central architects of the Vietnam War, cited the memory of the Munich Agreement as one of the reasons behind American involvement in Vietnam. President Lyndon Johnson even used the word *appeasement* to reject peace terms with North Vietnam. Lost in this collective memory is the fact that the Munich Agreement was extremely popular at the time of its signing because it held the promise of peace. Daladier returned to France deeply disturbed by a sense that he had failed not only Czechoslovakia but the interests of France. Yet to his astonishment, his return was met by crowds of cheering Frenchmen. The irony is that while most of the world was cheering the policy of appeasement, some of Chamberlain's harshest critics, along with the prowar lobby of Winston Churchill and Anthony Eden, were peace activists who understood that appeasement only served Churchill's argument for war. In 1938, Oswald Garrison Villard, an outspoken pacifist writing in *The Nation,* characterized Chamberlain's approach as "his shameful and stupid policy of keeping peace in Europe by wholesale surrenders to the dictators."

Once war broke out it became fashionable to say, as it is still said today, that World War II was caused by the policy of appeasement—that by not taking a firm stance against Hitler's seizure of Austria, the Sudentenland, and Czechoslovakia, Nazi Germany had been allowed to grow into an uncontrollable monster. What is seldom mentioned is the decade of support and acquiescence by political and business leaders prior to Munich. This was despite the fact that Hitler had made clear in the 1920s his intention to invade France, take Austria and Czechoslovakia, and destroy "inferior races," which he called *Untermenschen.*

In 1939, when Stalin signed a nonaggression pact with Hitler in an attempt to keep the Nazis expanding westward rather than eastward, the preponderance of Western thinking started turning antifascist. Antifascism was a temporary condition that only lasted through World War II. After the war, the Allies administered Germany with a policy called de-Nazification that was discontinued in the late 1940s because as the Cold War intensified, the Allies wanted to make use of the strong anti-Communist leanings of Nazis. Numerous former high-ranking Nazis were left to assume important roles in the rebuilding of what became West Germany.

———

A 1935 poll indicated that 75 percent of Americans were in favor of requiring a national referendum before going to war. When the same question was asked in 1939, only 59 percent were in favor of the referendum. In 1940, the officers at Fort Lewis, California, nicknamed General Dwight Eisenhower "Alarmist Ike" for insisting that the United States was going to war. Americans no longer wanted any part of European wars.

But three days after the Japanese attack on Pearl Harbor, a poll showed 96 percent approval for the congressional declaration of war. Jeanette Rankin, one of fifty Representatives who voted against the 1917 House declaration of war, was the only House member to vote against the 1941 declaration of war.

During the war a poll taken in the United States showed that 87 percent of Americans found the expression of pacifist ideas objectionable. Nevertheless, some held firm. Three times as many men had conscientious-objector status and four times as many went to prison rather than serve in the military in the United States during World War II as had during World War I. In the United States 42,973 men refused to fight in World War II.

Among them was Ralph DiGia, born in New York to Italian immigrants in 1914. His father considered himself "a radical antifascist." Starting at the age of ten, the year Mussolini secured control of Italy, Ralph was taken by his father to antifascist meetings. At thirteen he marched in his first demonstration.

In 1940 Ralph registered for the draft as a conscientious objector. "My father said, 'You are a college graduate. You can get an easy job in the Army and you can still keep your ideas .... Don't ruin your life,'" recalled DiGia, interviewed in his office at the War Resisters League at the age of ninety-one. "I never expected to stop the war," he said. "But you have to stand up for what you believe in or nothing ever changes." The Selective Service gave him a hearing and rejected his CO claim because he belonged to no church and had no religious background.

During World War II, one in every six inmates in federal prisons was a conscientious objector. While the fighting Allies remained silent about the Holocaust, some of the few who had been outspoken about the barbarism of the Fascists were languishing in American prison for refusing to fight the war. They took on the cause of prison racial integration, and their strikes to integrate prison mess halls were among the first acts of the twentieth-century American civil rights movement.

———

In India in November 1938 Mohandas Gandhi wrote of European Jewry: "I am convinced that if someone with courage and vision can arise among them to lead them in nonviolent action, the winter of their despair can in the twinkling of an eye be turned into the summer of hope." But the Jews of Europe, like the Indians and the black slaves in America, were not able to find an effective means of resistance. They met their fate either passively or with violent resistance, either of which responses resulted fairly quickly in their deaths. The Nazis are often cited as an example of an enemy against whom nonviolence would be futile. This is said despite the success of several nonviolent campaigns. Amid some of the greatest violence the world has ever seen, it was little noted that more Jews were saved by nonviolence than by violence.

Denmark was a neutral country trying to stay out of the war when the Germans, claiming the Allies were planning to force them out of neutrality, demanded Denmark's capitulation, and in return for not resisting they promised that Danish independence would be respected. Denmark, regarding armed resistance as suicidal, submitted passively to German occupation. It became a point of national honor to work slowly, delay transportation, destroy equipment, and, above all, to protect anyone the Germans pursued. Youths openly demonstrated against German policies. Underground groups sabotaged trains and other infrastructure. Workers went out on strike around the nation. Only one strike turned violent, in Odense, where the Germans fired into an angry crowd, wounding four, including a child, and the surging crowd beat to death the German soldier who had fired.

The Danish government had refused to enact any anti-Semitic measures, and on October 1, 1943, when the Germans announced their decision to deport Jews from Denmark, the Danes hid almost the entire Jewish population of 6,500, including about 1,500 refugees from Germany, Austria, and Czechoslovakia. The hidden Jews were then taken by boat to neutral Sweden. The Germans only succeeded in deporting four hundred to Theresienstadt. The Danish government relentlessly inquired on their behalf and at one point managed to send representatives to visit them. Because of this close attention by their government, no Danes were in the transports sent to Auschwitz. Fifty-one died of sickness. The rest of the Jewish population of Denmark survived. Compare this to France, which had one of the better records, where there was well-organized armed resistance but 26 percent of 350,000 Jews were lost; or the Netherlands, where three-quarters of a Jewish population of 140,000 were killed despite armed resistance; or Poland, where 90 percent of 3.3 million Jews were killed despite an armed Polish resistance and armed Jewish uprisings.

Of the Jews who were saved from deportation to concentration camps, very few were saved through violence. The government of Bulgaria, a German ally, saved their Jewish population by refusing to cooperate. Raoul Wallenberg, a Swedish businessman saved an estimated 100,000 Hungarian Jews while serving as his country's ambassador to Hungary by either issuing Swedish passports or moving Jews to secret locations. André and Magda Trocmé, a Protestant minister and his wife, who in the 1930s had established a school to study nonviolence in southeastern France, during the occupation organized most of the village of Le Chambon-sur-Lignon to hide Jewish children and transport them across the Swiss border. They saved several thousand children. World War II abounds with such tales of nonviolent resistance and noncooperation. Hundreds of thousands of Jews were saved by individuals who risked their entire family to hide a Jew or a Jewish family.

Dictatorships are prepared to crush armed resistance; it is noncooperation that confounds them. "What could he [the dictator] do to you," Montaigne's good friend Étienne de la Boétie asked in a 1548 essay on dictators, which was pointedly republished in the United States in 1942, "if you yourself did not connive with the thief who plunders you."

——

There was something odd about the war propaganda machine. Since hatred of the enemy is a cornerstone of selling a war, in World War I the British and American presses, in collusion with their governments, made up the most outlandish lies about German atrocities. The Kaiser was portrayed as monstrous, "a lunatic." German soldiers were said to rape nuns and mutilate children. H. G. Wells, who invented the phrase "the war that will end war," also invented the myth of "Frankenstein Germany," the monster state. A story broken by the *Times* of London, that Germany had a factory that turned corpses into munitions, was widely believed, though completely fabricated.

But in World War II, when Germany really was led by a lunatic, when Germans did mutilate and murder children, when they had death factories that actually did make soap out of human beings, little of this was included in the war propaganda. The governments of the Allied nations had not abandoned propaganda. And yet the Holocaust, the systematic murder of six million Jews, was a subject rarely touched upon in the media. Contrary to popular postwar claims, the Holocaust was not stopped by the war. In fact, it was started by it. Before the war, Jews had been stripped of their rights and property and in some cases thrown into labor camps along with Communists and political dissidents. Various schemes emerged, including one in 1940, shortly after the war had begun, to deport Jews to Madagascar, a plan that failed because it would have had to be negotiated with France and Britain and this could not be done in wartime. Only in the isolation and brutality of wartime, in 1941, after the invasion of the Soviet Union in late June, when Germans had millions of additional Eastern European Jews under their control, did Germany dare to turn concentration camps into death camps. And only in January 1942, at a secret conference in the Berlin suburb of Wannsee, did the Germans plan *Die Endlösung,* the "final solution," killing them all. In the postwar world it became fashionable to view the Allied military effort as an attempt to stop the Holocaust. But in reality the Allies went to war over geopolitical concerns. If they had wanted to save the Jews, the best chance would have been not going to war. But as with the slaves in the American South, there were too few interested in the plight of Europe's Jews.

In recent years, formerly secret documents have been released that make it clear that the Allied governments and militaries were well aware of the genocide in progress and consciously chose not to interfere with it. The claim often made by German citizens, by the populations of Britain and America including Jews, and by the British and American governments that they didn't know what was happening is simply not true. The very efficient code decipherers of British intelligence were reading reports from death camps at least as early as 1943. They may have intercepted messages earlier, but documents confirming this are still not available. But it is known that they listened in on conversations about the unsatisfactorily slow killing rate at Auschwitz. A Polish agent, Tadeusz Chciuk-Celt, parachuted into Poland in 1942 and reported back on the killing. Also in 1942 the Polish underground reported on the mass extermination of Jews. In 1943, the Polish government in exile in London urged the British to bomb Auschwitz, where Polish political prisoners were being killed. In any event, by 1944 escaped

Auschwitz inmates had told the world exactly what was taking place there.

According to Holocaust historian Martin Gilbert, representatives of the Jewish Agency for Palestine met in the summer of 1944 with Winston Churchill, at which time they spoke with him about the accelerated deportation and extermination of Hungarian Jews at Auschwitz and urged him to bomb the rail lines to stop the deportation. But nothing was done. Gerhart Riegner of the World Jewish Congress in Geneva, who first informed the Allies of the operations of the Final Solution in 1942, went to both the Americans and the British in 1944 at the request of Slovak and Hungarian Jews and asked them to bomb Auschwitz. He was told: "It cannot be done. It is too far away."

Studies of Allied reconnaissance photos of the I. G. Farben synthetic-oil plant, bombed four times in 1944 by the U.S. Fifteenth Air Force based in Italy, show Jewish arrivals being marched to crematoriums and even the crematoriums themselves at Auschwitz-Birkenau, a mere five miles away. In fact, on September 13, 1944, the Fifteenth did bomb Auschwitz-Birkenau, but by accident. According to Dino Brugioni, who flew reconnaissance missions over Europe and later worked for the CIA, by the time the Soviet army liberated Auschwitz, on January 27, 1945, and the world "learned" of its horrors, the Allies had photographed the Birkenau death camp from the air at least thirty times.

The Jewish Agency, Riegner, and others who pleaded for help to stop the Holocaust had been repeatedly told: "We have a war to win." For the Allies, stopping the Holocaust was militarily irrelevant, and from a purely strategic point of view this was probably true. But more to the point, neither Roosevelt, Churchill, nor most of all Stalin wanted to make the war about saving the Jews, because, as with freeing the slaves, going to war to save the Jews would not have been popular. The many anti-Semites in the United States, Britain, and France would have been, or at least it was supposed that they would have been, resentful of being asked to fight for Jews. Nazi propaganda minister Joseph Goebbels repeatedly claimed that the Allies were attacking Germany because they were controlled by Jews. Churchill and Roosevelt understood the potency of this claim and did not want to give it credence. Roosevelt had been criticized sharply after the 1936 election, when he slightly opened up Jewish immigration so that of the 300,000 Jewish refugees taken in by the world, who were a mere fraction of those trying to escape, two-thirds were received by the United States. This led to accusations that Roosevelt was "too close to the Jews," or that he was being manipulated by them.

The Allies were so sensitive to these accusations that they even rejected several requests to simply announce that bombing raids that they were carrying out on German cities, in any event, were being carried out as vengeance for the death camps. Since the argument for these bombings of civilian targets was to destroy German morale, it was thought by some, including General Sikorski of the exiled Polish government, that it would have a demoralizing effect on Germans to know that their cities were being bombed because of Hitler's Final Solution.

————

Ever since the outset of the war, books and films have been using World War II to promote the concept of just war. A typical example is the 1998 Steven Spielberg movie *Saving Private Ryan*, which focuses on the 1944 D-Day invasion. This film is sometimes said to be antiwar because it uses a great deal of noise and special effects to make war look gory, though it does not make it look nearly as grim as Lewis Milestone's 1930 black-and-white adaptation of *All Quiet on the Western Front*, which had no special effects at all. Yet the 1998 World War II film, in sharp contrast to the 1930 World War I film, gives the message that this terrible sacrifice is "worth it," which is the central lie in the promotion of warfare. Spielberg's saga, in a flourish worthy of Urban II, even argues that ignoring the Geneva Conventions and murdering prisoners of war is a reasonable act since the enemy is so insidious.

The same year that *Saving Private Ryan* was released, news broadcaster Tom Brokaw came out with a fairly typical example of post–World War II militarist propaganda, a book titled, without the least embarrassment, *The Greatest Generation.* According to this book, this "greatest generation," which first failed to oppose fascism and then repeated the slaughter of 1914 with even more ruthlessness, had "answered the call to save the world."

Such orgasmic rhetoric has its roots in the war effort itself. In Franklin Roosevelt and Winston Churchill the United States and Britain had talented leaders given to patrician eloquence. These were leaders who were, as Tolstoy wrote of Napoleon in *War and Peace,* "capable of justifying, in his own name and in theirs, all the duplicity, robbery and murder that ensued." Churchill in particular was a master of Urban II–ship. The cause was holy, the enemy was Satan: Churchill's entire career was spent in the promotion of warfare. He was part of the World War I British government, one of the war promoters whose deceptions were denounced after the war.

But the claim that a just war was being waged in order to stop the Holocaust did not come until years later. The Allies were simply good; the Germans and Japanese bad; and those who fought for the Allies were heroes. In truth the "greatest generation," like all men who go to war, were simply doing what they were told, exactly like the men of the lesser generation before them.

The difference was that the "heroes" of World War II were not allowed to talk about such things, were not allowed to air the many things troubling them, or even to admit to the guilt they felt. Scarcely any of the many World War II movies ever admitted that these men suffered psychological damage. Two of the rare exceptions were William Wyler's *The Best Years of Our Lives,* made right after the war, in 1946, and Nunnally Johnson's *The Man in the Gray Flannel Suit,* which appeared ten years later, in 1956. The trauma of World War II veterans was so repressed that many did not even realize they were suffering until they started learning of the "syndromes" of Vietnam War veterans and noticed that much of it sounded very familiar.

———

War resister Ralph DiGia said, "World War II reinforced my belief that in war one becomes what the enemy is accused of being." The Allies committed atrocities that would have appalled them if committed by anyone else. The outrage with which the British and the

Americans responded to the 1937 German bombardment of the Basque town of Gernika shows the extent to which those two populations regarded the deliberate bombing of civilian targets as an indefensible crime. Ever since a German zeppelin attack on London killed 127 people in October 1914, the British political leadership had been outspoken on the immorality of bombing civilians.

Carl von Clausewitz, a Prussian general who had fought against Napoleon, maintained in his classic nineteenth-century text *On War* that war "theoretically can have no limits"; but the British and the Americans, and most other people in the world, thought it did; putting limits on it was what made it acceptable, and targeting civilians, especially with the new and deadly weapon of aerial bombardment, was thought to be an unacceptable excess. But the problem with war, as former Allied commander, later President Dwight Eisenhower, observed in a 1955 press conference, is that once you start, you get "deeper and deeper," until the only limitation is "force itself." Which was Ralph DiGia's point.

Allied bombing of German cities during World War II killed approximately 300,000 German civilians and wounded almost 800,000. It is an emotionally moving though legally and morally irrelevant argument that these are small numbers for a people who exterminated six million in death camps. Of course, at the time, that argument was not even being made. In late 1940 the Germans bombed Coventry from seven in the evening until past six the following morning. Six hundred residents and a fourteenth-century cathedral were destroyed. Now British Bomber Command was given instructions to retaliate on German cities, and the targets would be the center of the city. In 1942 they were specifically told not to aim for military targets, such as aircraft factories and shipyards, but at densely populated areas, since the goal was to destroy the population's morale, which had also been Germany's goal in Coventry. With demoralization in mind, working-class districts in particular were to be sought out.

What is surprising is that the British government does not appear to have noticed that the Germans had failed to damage morale in Coventry or London. On the contrary, those bombings seemed to rally support among the British population for the war effort, although they did not rally support for retaliatory civilian bombings. Churchill presented the British raids as vengeance for German bombing, but opinion polls showed that the raids were much more popular in unaffected areas of England than among bombed populations such as London. War is always more popular with those who don't experience it. In fact, polls also showed that the raids remained popular only because most of the population insisted on believing that they were really against military targets.

Some in the military objected. Rear Admiral L. H. K. Hamilton protested, "We are a hopelessly unmilitary nation to imagine that we win the war by bombing German women and children instead of defeating their army and navy." Some historians believe targeting civilians took away from military targets and may have actually prolonged the war. The bombing, with no military purpose, continued to the very end of the war, and the Germans fought on until their military was completely routed, without civilian morale ever becoming a factor.

In February 1945, when Germany was near surrender, Dresden was firebombed in a joint British-American attack in a way clearly designed to destroy the baroque center of the historic city. Historians estimate that between 100,000 and 130,000 people—a third of total German civilian casualties—died in that bombing alone. The fact that in so doing the Allies had destroyed Gestapo headquarters, stopping a deportation convoy the next day that would have meant the death of the remaining 198 Jews from a population of 5,000, was a mere coincidence of no relevance to Allied plans.

The killing of German civilians led the way for far worse slaughters of Japanese civilians by the Americans. Shortly after the Dresden success, the Americans had firebombed Tokyo with a similar result of about 100,000 civilians killed. Before the American atom bomb was completed, it was learned that the alleged reason for the American atomic bomb project, the German program to build an atomic bomb, had failed. This news did not slow down the building of America's nuclear weapons, which could now be used against the Japanese, even though Japan had no nuclear program. In justifying the attack, Truman cited Pearl Harbor, where the United States had been attacked "without warning." What he did not mention is that Pearl Harbor was an entirely military target, whereas in Hiroshima and Nagasaki, civilians were the target. His other explanation, that the bombs' purpose was "to shorten the agony of war," was a thin justification, but one that Churchill immediately joined in on, saying that "to avert a vast, indefinite butchery . . . at the cost of a few explosions seemed, after all our toils and perils, a miracle of deliverance."

This ignores the fact that Japan had been ready to surrender before the nuclear attacks but the United States insisted, as they had with Germany, that the surrender be unconditional. The Japanese wanted to negotiate. Those negotiations would have saved not only the enormous losses from a land invasion of Japan but the estimated 120,000 civilians who were killed from those "few explosions" and an equal number wounded, many horribly disfigured, and most of whom eventually died.

Perhaps the real reason for the bombing is revealed in the U.S. government publication *United States Strategic Bombing Survey, Summary Report on the Pacific War,* dated July 1, 1946:

> On 6 August and 9 August 1945, the first two atomic bombs to be used for military purposes were dropped on Hiroshima and Nagasaki respectively. One hundred thousand people were killed, six square miles or over fifty percent of the built-up areas of the two cities were destroyed. The first and crucial question about the atomic bomb thus was answered practically and conclusively; atomic energy had been mastered for military purposes and the overwhelming scale of its possibilities had been demonstrated.

Or perhaps Eisenhower had it right when he observed that once you start the business of killing, you just get "deeper and deeper." What would that now mean for a world with nuclear weapons?

X

THE RULE OF THUGS AND THE LAW OF GRAVITY

Whether mankind will consciously follow the law of love, I do not know. But that need not perturb us. The law will work, just as the law of gravitation will work whether we accept it or no. And just as a scientist will work wonders out of various applications of the laws of nature, even so a man who applies the law of love with scientific precision can work greater wonders.

—MOHANDAS GANDHI,
published in *Young India*,
October 1, 1931

Since the close of the twentieth century it has become common-place to refer to it as the most catastrophically bloody century in history. Lenin, who saw war as "an inevitable stage of capital-ism," had predicted this at the century's beginning. By the end of the century, an estimated 187 million people had died in war, the equivalent of 10 percent of the planet's population at the outbreak of World War I. That made it a record century, but also a far higher percentage of war fatalities were civilian than in any previous cen-tury. In World War I, one-fifth of casualties were civilian, but in World War II it went up to two-thirds. In twenty-first-century war-fare, such as in Iraq, the casualties may be as high as 90 percent civilian.

But it is seldom mentioned that the twentieth century was also the greatest century for nonviolent activism: from 1945 until the close of the century the world saw more victories for nonviolence than ever before in history. It started with a peculiar man in India. Mohandas Karamchand Gandhi, called Mahatma, "the great soul," was an example of how a rebel genius was killed and then turned into a saint, the easier to ignore his legacy. For Gandhi did not want a militarized India with a nuclear weapons program.

Gandhi was a quirky man with a mischievous sense of humor. Asked what he thought of Western civilization, he replied, "I think it would be a great idea." But he was not a narrow nationalist. He was a British-educated lawyer influenced not only by his own Hin-duism but also by Jainism, Buddhism, and the teachings of Jesus. He was a great admirer of Leo Tolstoy, who had withdrawn to his country estate to rail against Christian clergy for denying the true teachings of Jesus. "Thus it is that these nations have become at-tached to a false Christianity," Tolstoy wrote, "represented by the Church, whose principles differ from those of paganism only by a lack of sincerity." Gandhi and Tolstoy corresponded with each other. Gandhi was also an admirer of Henry David Thoreau, whose small book *Civil Disobedience,* written while the author was in prison, Gandhi had read while himself in prison.

Gandhi was attracted to eccentric theories of sex, diet, and bodily functions. He had lusts and passions that infuriated him. A tiny man, as a youth he defied his parents' interdiction against the eating of meat, hoping that a meat diet would make him large and strong like the carnivorous British. His marriage had been arranged when he was thirteen and lasted until his death, sixty-two years later. No doubt a good Freudian would have liked to have had him on the couch. At sixteen, his father died while Gandhi was having sex with his young wife. He later said, "This shame of carnal desire . . . is a blot I have never been able to efface or forgive." This seems to be at the root of an impulse to deny physical pleasure. Into old age he convinced attractive young women to lie naked through the night with him in order to test his resolve to remain chaste. He also lived much of his life without money or possessions.

Despite frequent expressions of inward doubt, he always displayed a confidence that allowed him to take unpredictable and unfashionable positions. He was fond of a quote from Thoreau: "The only obligation which I have a right to assume is to do any time what I think right." He supported the British in World War I despite his absolute rejection of warfare—a position that left his followers confused. Gandhi believed that to defeat an enemy, the enemy must not feel defeated or humiliated. He did not want the British to be bitter, as they would have been if India turned against them in the World War.

He had no doubts, however, about the power of nonviolence. In 1921 he wrote: "Given a just cause, capacity for endless suffering, and avoidance of violence, victory is certain." This confidence reassured his followers and unsettled his adversaries, though neither completely understood him.

The British tended not to take him seriously. Like the American revolutionaries, he advocated the making of homespun and rejected use of cloth from British mills. To the British this made for a comic figure, the spindly little man dressed in beggar's rags. Many British leaders, including Winston Churchill, spoke scornfully of him. When the British colonial government finally and reluctantly recognized that he was a man to be reckoned with and agreed to meet with him in 1931, Churchill angrily snorted about "the nauseating and humiliating spectacle of this one-time Inner Temple lawyer, now seditious fakir, striding half naked up the steps of the Viceroy's palace, there to negotiate and to parley on equal terms with the representative of the King-Emperor."

Gandhi wrote to Churchill: "I would love to be a naked fakir but am not one yet." A fakir is an Indian monk who wanders, begging for sustenance.

But the real Gandhi, unlike the saint, was not a dreamer—he was a tough pragmatist who focused on winning. As he pointed out, "Strength does not come from physical capacity, it comes from an indomitable will." Churchill soon understood that this combination of nonviolence and pragmatism made Gandhi a dangerous opponent. In 1935 he said, "Gandhism and all that it stands for must finally be grappled with and crushed."

Gandhi was first and foremost a political activist, and he had utter contempt for nonactive pacifism. Like William Lloyd Garrison, he regarded such a passive stance as cowardly, calling inaction "rank cowardice and unmanly," and said that he would rather see someone incapable of nonviolence resist violently than not resist at all. "Violence is any day preferable to impotence," he wrote. "There is hope for a violent man to become nonviolent. There is no such hope for the impotent." Feeling so strongly about the distinction between nonviolent resistance and pacifism, he was dissatisfied with the absence of a proactive word for his beliefs. *Ahimsa*, nonviolence, the absence of violence, did not begin to express the active nature and strength of a program of political action. And so he created the word *satyagraha*, literally "truth force."

He began his first *satyagraha* campaign as a thirty-five-year-old lawyer in South Africa and after seven years got the South African Indian government to agree to end discrimination against Indians. Black Africans have frequently criticized him for having failed to show any interest in the plight of the black South African majority. But Gandhi was unpredictable. He expected his loyalty to Britain during the war, a gesture few understood, to be repaid with moves toward independence after the war, but the British appeared unmoved. In the 1920s he began campaigning for Indian independence with strikes, boycotts, and protests.

It is notable how much his program to expel the British resembled the American program before it turned violent in 1775. Like a twentieth-century John Adams, Gandhi wrote in 1930:

> Much can be done. . . . Liquor and foreign cloth shops can be picketed. We can refuse to pay taxes if we have the requisite strength. The lawyers can give up practice. The public can boycott the Courts by refraining from litigation. Government servants can all resign their posts. . . .

The argument is frequently made that Gandhi was able to succeed only because of the gentle nature of his opponents. Religious philosopher Reinhold Niebuhr, for one, in his long-standing disagreements with Abraham Johannes Muste, a Calvinist minister turned labor organizer turned peace activist, argued, "Pacifism was irrelevant in dealing with Hitler." There are several problems with this argument, the first being that the Danes and isolated groups of religious pacifists in other countries had demonstrated that even against Nazis nonviolence could achieve some goals. But those who dismiss Gandhi's accomplishments because they were "only against the British" are also overlooking how ruthless and brutal British colonial rule could be. The history of British rule on the Subcontinent belies this myth, especially their treatment of the Pathans along the Hindu Kush, with its strategic Khyber Pass, where the British tried to control by fear the gateway from Afghanistan to India for a century.

In 1842 the British attempted to secure the area by sending their 4,500-man Army of the Indus through the Khyber Pass. One survivor made it to Fort Jalalabad. But the British were determined to subdue the Muslim tribesmen, the Pathans, who were said to be one of the most warlike people in the world. So were—it is so easily forgotten—the British.

The British sent expedition after expedition into the Pathan hills, an area known to the British colonial army around the world as "the grim." In the nineteenth century, they shelled the Pathan villages. In the twentieth they bombed them from the air. Thousands of Pathans were flogged or otherwise beaten. But Pathan snipers fired ancient, handmade rifles from somewhere in the rocky crests. Then in the 1930s something happened that made them more dangerous, more threatening than the British army had ever imagined: the Pathans joined up with Gandhi's nonviolence movement. The British knowingly asserted that this was a trick, that Pathans were not capable of nonviolence. They sealed off the area and for two decades brutalized the nonviolent resisters. Pathans were shot down in large groups, tortured, jailed, flogged, imprisoned for life in distant Indian Ocean penal colonies, or hanged. The British army burned their fields and their houses. There was no due process of law in the Pathan zone.

Their leader, Abdul Ghaffar Khan, called Badshah Khan, the Khan of Khans, was almost the complete opposite of Gandhi, the gentle Hindu. A photograph of the two of them appeared to be manipulated, for Badshah Khan, the Pathan aristocrat, was a mountainous man with broad shoulders and a square, strong-featured face surrounded by the thick hair of his head and beard. Gandhi, from the humble side of a middle-class caste, barely came up to his shoulders and was frail, and his baldness gave roundness to his small head. But they were firm allies, determined to build, through nonviolence, an independent India for Muslims and Hindus together.

In 1929, a young man who had heard Khan speak gave him the idea of organizing in a way that was consistent with Pathan tradition, and so he recruited the world's first nonviolent army, the Khudai Khidmatgars, the Servants of God. Any Pathan could join Khan's army by swearing an oath to renounce violence and vengeance, to forgive oppressors, and to embrace a simple life. Khan quickly recruited five hundred soldiers who opened schools and maintained order at gatherings and demonstrations. Khan went from town to town urging Pathans to rise up in civil disobedience.

In Peshawar, when Khan was arrested by the British, the entire town's population took the oath and joined his army. The region was stopped by a general strike and the British sent in the army with armored vehicles. When they began firing into the crowd, the demonstrators stood stoically. Some were shot many times. One boy walked up to a soldier and asked him to shoot him, and the soldier shot the boy dead. As people fell, others moved forward to be shot. The British continued shooting for six hours and then began the work of hauling away the bodies and burning them. The result was that 80,000 new volunteers took the oath and joined Khan's army.

One platoon of an elite Indian regiment, the Garhwal Rifles, refused to fire and every man in the outfit was sentenced to stiff prison terms, one for life; and even when negotiations forced the British to release political prisoners, all the Garhwalis were made to serve their full terms.

The British tried to provoke the proud Pathan soldiers into breaking their vows of nonviolence. Huge, powerfully built men were publically stripped, humiliated, beaten with rifle butts, poked with bayonets, thrown in cesspools. Some killed themselves to avoid breaking their vow.

In prison, Badshah Khan explained to his jailer that he was a nonviolent follower of Gandhi. The British deputy commissioner asked him what he would be doing if he hadn't met Gandhi and Khan put his large hands on the two bars in front of him and easily bent them apart.

Khan's stated concept of Islam was this:

The Holy Prophet Mohammed came into this world and taught us, "That man is a Muslim who never hurts anyone by word or deed, but who works for the benefit and happiness of God's creatures. Belief in God is to love one's fellow men."

For putting these dangerous ideas into practice, the "gentle British" held him for a total of thirty years, one third of his life, serving various prison terms, most of it in solitary confinement, usually under a charge of "sedition."

———

When World War II broke out Gandhi did not want to repeat his mistake of the first World War. This time he offered to support the British only if they would promise independence after the war. They declined, and he continued his campaign, for which he spent two years in prison. After the war, facing further disruption, the British agreed to independence in 1947.

But the terms of independence were a disappointment to Gandhi, since they divided the Muslims and Hindus into Pakistan and India. This is the issue over which he was assassinated in January 1948 by a fanatical Hindu who feared that his nonviolent approach would give too much away to the Muslims. While Gandhi is remembered by the nation he created, his teachings are not followed. His closest collaborator, Jawaharlal Nehru, leading the newly independent India, quickly responded to challenges from Pakistan on the border, China from Tibet, and Portugal in Goa, by building a militarized Indian state. But the enduring importance of Gandhi in the world was that he demonstrated in pragmatic steps how nonviolence could work. Nowhere did he have a greater impact than in the United States.

———

In the 1920s a Quaker lawyer, Richard Gregg, stumbled across Gandhi's writings by chance. In 1925, having studied everything he could find by Gandhi, he decided to follow him to India, where he spent the next four years, including seven months at Gandhi's ashram. Returning to the United States in 1929, he gave up his law practice and instead wrote and lectured on Gandhi's theories of nonviolence. He became the leading American theorist of nonviolence, publishing *The Power of Nonviolence* in 1934, a book he dedicated to Gandhi. For those accustomed to reading on nonviolence, this book had an oddly pragmatic tenor. His main point was that nonviolence works, that it is an effective way to get things done.

George Houser, along with David Dellinger, had been one of nine students at the Union Theological Seminary in New York City who refused to register for the draft in 1940 and were sent to prison. Houser had written to A. J. Muste (who *Time* magazine in 1937 called "America's number one pacifist"), and their correspondence continued when Houser was in prison. Much of the dialogue was about the importance of establishing an official nonviolent movement. As a leader in the United Textile Workers Union, Muste, himself much influenced by the writings of Richard Gregg, had moved the union toward nonviolent tactics, starting with a 1936 "lie-down picketing." In 1937, Alexander McKeown, the national vice president of the American Federation of Full-Fashioned Hosiery Workers, said:

> As a rule men and women hesitate to adopt the tactics of a Gandhi in an industrial or civil dispute for fear of seeming to make fools of themselves. The fact of the matter is that nonviolence is a tactic that requires perhaps a higher type of courage and devotion than is called for in ordinary physical combat.

In 1941 Muste became director of the Fellowship of Reconciliation (FOR), which had been founded in 1915 by anti–World War I protesters and was by then run by traditional pacifists, followers of what Muste called "the sentimental, easygoing pacifism of the earlier part of the century." Muste, like Gandhi, saw pacifism as a tool for political activism. "In a world built on violence," he wrote in a 1928 essay, "one must be a revolutionary before one can be a pacifist."

With World War II in progress, it was difficult to attract support for nonviolence in international affairs, but Muste believed that the techniques of nonviolent resistance would prove particularly effective in race issues, such as the desegregation of the South. In seeking a new young staff for the FOR, he brought in James Farmer, a large man with a booming voice who had recently received a doctorate from Howard University, where he had studied the teachings of Gandhi. Farmer was a Southerner, the son of a college professor. His father, the son of slaves, on receiving a Ph.D. from Boston University in 1918, became one of only twenty-five African-Americans ever to hold a Ph.D. Muste also brought in Bayard Rustin, a young black Quaker from Pennsylvania who had recently, like Muste, abandoned the Communist left for a new kind of radicalism. Rustin was a tall, handsome young man with a rich tenor voice who, prophetically, had recorded the song "Chain Gang" with the folk/blues great Josh White.

Farmer and Rustin, along with George Houser, became active in an organization formed in Chicago in 1942 called the Congress of Racial Equality (CORE). CORE had a guru-like adviser, Krishnalal Shridharani, a disciple of Gandhi despite his love of women, food, cigars, and other earthly pleasures. "Just where everything will lead, we do not know," Houser wrote to Muste, informing him of the creation of CORE. "But sooner or later we are likely to hit on something big." In 1943, sit-ins forced the integration of Denver movie theaters and integrated an all-white cafeteria in Detroit.

Rustin, Houser, and Farmer were all World War II resisters. In 1944 Rustin denounced his Quaker draft deferment and went to prison for twenty-seven months to protest the war.

In 1947 Houser and Rustin organized the "Journey for Reconciliation," testing a recent Supreme Court decision that found segregated interstate travel accommodations to be unconstitutional. This is considered the first "Freedom Ride," except for a 1942 bus trip from Louisville to Nashville when Rustin, traveling alone, refused to sit in the back of the bus. He was taken off the bus and taken to a local police station, where he was severely beaten. Following the Gandhian principles of resistance, Rustin did not fight back but kept trying to reason with the men as they worked him over. Finally one of the white passengers interceded for his release. "I believe the nigger's crazy," said the police chief.

Richard Gregg called it "moral jiujitsu." The attacker expects resistance, and when there is none he loses his "moral balance." It was effective but dangerous. James Farmer, who had endured numerous beatings, said in a 1991 interview, "Anyone who said he wasn't afraid during the civil rights movement was either a liar or without imagination."

In 1947 Houser and Rustin recruited eight whites and eight blacks to sit in the "wrong" section of segregated buses on a two-week journey through Virginia, North Carolina, Tennessee, and Kentucky. The head of the NAACP's legal department, Thurgood Marshall, warned against the plan, saying that "a disobedience movement on the part of Negroes and their white allies, if employed in the South, would result in wholesale slaughter with no good achieved."

Bayard Rustin and another black, Andrew Johnson, were sentenced in North Carolina to thirty days on a chain gang. Rustin, who had learned his Gandhi well, gladly accepted his sentence, saying that it might be good for the cause. When he got out, he went to India for six months at the invitation of the ruling Congress Party.

Rustin went on to gladly accept more beatings. Marching against the Korean War in 1951, he was attacked with a stick by an angry spectator. Rustin handed him a second stick and asked him if he wanted to use both. The attacker threw both sticks down.

Rustin once said of his mentor in nonviolence, A. J. Muste, "During all my work with Martin King, I never made a difficult decision without talking the problem over with A.J. first." Rustin and Farmer were not the only civil rights leaders influenced by Muste. In 1949, Martin Luther King Jr., a young student at Crozer Theological Seminary, in Chester, Pennsylvania, who had written on his application that he had "an inescapable urge to serve society," attended a lecture by Muste on nonviolence. Francis Stewart, a white friend from Georgia, later recalled "a pretty heated argument" between Muste and King after the lecture. "King sure as hell wasn't any pacifist then." King himself stated that most of the time he was in the seminary he continued to believe "the only way we could solve our problem of segregation was an armed revolt." Later that year, King attended a lecture about the work of Gandhi by Howard University president Mordecai Johnson, freshly returned from a trip to India. And after he had changed his thinking, he talked about his first encounter with Muste:

> I wasn't a pacifist then, but the power of A.J.'s sincerity and his hard-headed ability to defend his position stayed with me through the years. Later, I got to know him better, and I would say unequivocally that the current emphasis on nonviolent direct action in the race relations field is due more to A.J. than to anyone else in the country.

In 1955, at the beginning of the Montgomery, Alabama, bus boycott, one of the actions that put King at the head of the civil rights movement, the FOR sent Bayard Rustin to instruct King in nonviolent techniques. King was immediately drawn to the older, more experienced Rustin as a mentor, but Rustin, while recognizing the young man's enormous talent, saw that he understood nothing of how to mount a nonviolent campaign. His house was strewn with firearms. Rustin laughed about nearly sitting on one by mistake.

The civil rights movement of the 1950s and 1960s became one of the most influential nonviolent movements in history, admired and emulated around the world. It helped shape an international anti-nuclear movement in which Muste, Rustin, and others from the civil rights movement were also deeply involved.

Nuclear weapons might have completely ended warfare. They ruined the concept of total war. Clausewitz had defined a victor as the side that ended up with the ability to impose its will, its program, on the other. A nuclear war would have no victor, because, as Clausewitz also pointed out, "War is a constant case of reciprocal action, the effects of which are mutual." An aerial attack like Coventry or Dresden might gain acceptance in a future war, and has, but not mutual Hiroshima, and the technology rapidly progressed to the point where nuclear weapons were many times worse than that. In 1969, political theorist Hannah Arendt wrote:

> The technical development of the implements of violence has now reached the point where no political goal could conceivably correspond to their destructive potential or justify their actual use in armed

conflict. Hence, warfare—from time immemorial the final merciless arbiter in international disputes—has lost much of its effectiveness and nearly all its glamour. The "apocalyptic" chess game between the superpowers, that is, between those that move on the highest plane of our civilization, is being played according to the rule "if either 'wins' it is the end of both"; it is a game that bears no resemblance to whatever war games preceded it. Its "rational" goal is deterrence, not victory, and the arms race, no longer a preparation for war, can now be justified only on the grounds that more and more deterrence is the best guarantee of peace. To the question how shall we ever extricate ourselves from the obvious insanity of this position, there is no answer.

Nuclear weapons did not end warfare, because that would have required political leaders to completely rethink their concept of power. As Hungarian writer György Konrád pointed out, the political elite had no alternative concept. "They have none because they are professionals of power. Why should they choose values that are in direct opposition to physical force?" And so, faced with nuclear destruction, the goal, rather than ending war, became limiting it.

Total warfare, using all available means to win, was no longer a viable concept. Words such as *glory* and *honor*, which David Low Dodge had called an "empty bubble" and a "standard of right and wrong without form or dimension," were at last stripped from the vocabulary. Clive Bell had written during World War I: "If we were sure that we could 'smash' the Germans only by smashing everyone else, ourselves included, I suppose we should desist from our endeavor." And that was exactly what war had become in the nuclear age.

Leading scientists, writers, political activists, and the general population around the world agreed on their opposition to nuclear weapons. The so-called "father of the Soviet hydrogen bomb," Soviet physicist Andrei Sakharov, and Albert Einstein, the Western "nuclear father," were in agreement. Sakharov took on Clausewitz's famous dictum that war was "a continuation of politics by other means," saying, "A thermonuclear war cannot be considered a continuation of politics by other means. It would be a means of universal suicide." While governments were playing out their Cold War, young people in both blocs were having their ideas shaped as they were being taught to crouch under their school desks, which someone somewhere had decided would be the one safe place for them in the event of a nuclear World War III.

Some films, such as Stanley Kubrick's 1964 *Dr. Strangelove, or: How I Learned to Stop Worrying and Love the Bomb,* and Stanley Kramer's 1959 *On the Beach,* as well as Nevil Shute's 1957 novel on which the Kramer film was based, presented the antinuclear argument, while a barrage of westerns of the same period taught that it wasn't good to fight but that when the bad guy comes to town, a man has to pick up a gun.

Throughout much of the world, the antinuclear movement steadily grew through the 1950s, so that a small, marginal, antinuclear group, SANE, founded in 1957, could three years later pack Madison Square Garden with more than 20,000 supporters.

A. J. Muste, borrowing an idea from Thoreau, refused to pay his taxes starting on January 1, 1948, in protest of the U.S. nuclear program. But unlike the nineteenth-century New England dissident, Muste was not immediately thrown in prison. Each year he sent the Internal Revenue Service a letter explaining why he was neither filing nor paying. The IRS did not even respond the first three years, and did not charge him until 1960, at which point they said he owed $1,165, plus penalties. Finally the court ruled that he could not be charged the penalties as he was following his conscience, though he did owe the taxes. But they had no way of collecting, since Muste owned nothing and did not even have a bank account.

The generation raised on the Cold War, rather than World War II, saw the world very differently than their parents did. When youth movements of the late 1960s all over the world were asked about the road that led them to rebellion they often mentioned the week in 1962 when the superpowers played that "apocalyptic chess game." Youth in both countries watched in horror as the exciting young president John F. Kennedy played moves and bluffs against

Nikita Khrushchev, the Soviet reformer responsible for sweeping away Stalinism. It was called "nuclear brinkmanship," and the two held everyone's lives like pawns.

While the conventional thinking of the World War II generation was that their war had secured the peace, albeit at a tremendous cost, those who grew up in the postwar period understood that there was no peace, that World War II had simply laid the groundwork for the Cold War, which was an umbrella term for more than one hundred shooting wars between 1945 and 1989. In the early 1980s, György Konrád, a Jew who had survived the Nazis in Hungary, counted approximately 130 postwar wars. "To find the main reason for today's threat of war," he asserted, "we must go back to the year 1945, to Yalta. . . . What a dirty trick of history! The Allies who were defending mankind from fascist inhumanity hastened, on the very eve of victory, to strike an imperialist bargain, a pact between Anglo-Saxons and Soviet imperialism."

The first global generation was coming of age more antiwar (and more anti–cold war) than their predecessors, but they were notably less nonviolent. This first became clear in the civil rights movement. In 1965, the Student Nonviolent Coordinating Committee (SNCC) was openly critical of Martin Luther King Jr. and his Southern Christian Leadership Conference. SNCC had been organized in 1960 to bring in students for the purpose of carrying out more sit-ins after the success of a lunch-counter sit-in in Greensboro, North Carolina. The organization later became involved in efforts to register black voters in the Mississippi Delta. By 1965, after enduring countless beatings and imprisonments, and numerous white and black volunteers killed, young SNCC activists started feeling that the charismatic Dr. King, whom they sometimes jokingly referred to as "de Lawd," was taking all the bows while they had all the ideas and took the punishment.

This was at a time of growing bitterness about the progress of the civil rights movement. SNCC took representatives of the newly registered black population of Mississippi to the Democratic Convention in 1964; but the Democrats, fearing an alienation of white

Southern voters, refused to seat them. Instead they alienated an important part of the civil rights movement.

SNCC, no longer "student" or "nonviolent," began emphasizing Black Power: self-reliance and the principle of meeting violence with violence. In 1966, the year Stokely Carmichael, a leading advocate of Black Power, took over SNCC, the Black Panther Party, which also advocated violence, was founded. The Black Panthers, like many a nonviolent activist, had an intuitive sense of street theater, but they used it to project a menacing violence. They posed dressed in black, with weapons, and used slogans such as "Off the pig!"—kill policemen. King's speeches were now sometimes booed by black activists or shouted down with cries of "Black Power!" In 1967, King said, "I'll still preach nonviolence with all my might, but I'm afraid it will fall on deaf ears."

The student anti–Vietnam War movement was also growing weary of taking beatings and was losing enthusiasm for nonviolence. Students had become fascinated by the anticolonialist liberation movements of the world—the Algerians and the Vietnamese, among others. One of the most popular books on college campuses in the late 1960s was Frantz Fanon's *The Wretched of the Earth,* originally published in French in 1961. Fanon was a psychiatrist from Martinique who studied in Algeria and became enamored of that country's fight for independence. Fanon at points appeared to glorify violence as the romantic and necessary element of the independence struggle. "Decolonization is always a violent phenomenon," Fanon wrote in the often-quoted opening paragraph of his book. He viewed colonization as a destructive violence and stated that "violence like Achilles' lance can heal the wounds it has inflicted." Hannah Arendt suggested that the problem was that students did not read past the first chapter, which is on violence. That is an exaggeration, since in the following chapter he writes, "Violence committed by the people . . . makes it possible for the masses to understand social truths." But he also warned that hatred and revenge "could not sustain a war of liberation" and that "unmixed and total brutality if not immediately combated, invariably leads to the defeat of the movement within a few weeks." That is exactly what happened to the Students for a Democratic Society, SDS, in the early 1970s.

SDS had been a leading force in organizing campuses against the war in Vietnam throughout the 1960s and at its height toward the end of the decade had 100,000 members around the country. To explore a linguistic absurdity, while not nonviolent, they were "not violent." Tom Hayden, a central figure in SDS during most of its ten-year history, said in a recent interview: "Nonviolence as a tactic is always to be preferred, but violence as either a threat or a real thing becomes inevitable in certain oppressive situations and has contributed to political or social change." David Dellinger, one of the principal organizers of the anti–Vietnam War movement, worked with Hayden and was somewhat distrustful of his commitment to nonviolence especially after the beating they took from Chicago police at the Democratic National Convention in the summer of 1968. As Stokely Carmichael observed decades later, nonviolence is "a very stern discipline."

Despite the participation of some sincere pacifists, such as Dellinger, the anti–Vietnam War movement was not specifically an antiwar movement. Many, probably a majority, of the protesters accepted war as a legitimate way to conduct business but felt that the motives of the United States in this particular war were immoral and that it was not a "just war." Some protesters openly supported the North Vietnamese side, which was certainly not nonviolent, apart from a minuscule nonviolence movement led by Buddhist monks that had failed to attract popular support in Vietnam or from the Western "antiwar movement." But SDS did refrain from violence even when attacked, until the group was taken over by the so-called Weatherpeople faction, a name change from the original politically incorrect Weathermen. This group, under the slogan "Bring the War Home," planned to terrorize the corporate establishment with a series of bombings. They had a change of heart after three of their own died in an accidental explosion, and, by design, they never hurt anyone in a series of late-night bombings.

Dellinger thought these young activists ranged from sincere but misguided to those with "deep-seated neurotic drives to violence. It would not be hard to imagine," he wrote, "that in a slightly different setting they could have been gauleiters, colonial police, or Green Berets."

Mark Rudd, one of the Weathermen founders, had led an essentially nonviolent movement that occupied five buildings on the campus of Columbia University in 1968. He began to contemplate violence after his group was attacked and beaten by a thousand rampaging New York City police. "The thinking was classic Marxism," said Rudd. "That the ruling class would never yield power without a fight. Also the act of fighting was itself liberatory, as in the writings of Frantz Fanon." At the time, police were assassinating Black Panthers, students were being beaten on campuses, and U.S. military forces were killing more than five thousand Vietnamese every week.

There is more than one lesson to be learned from the SDS experience. First, there is the lesson that violence always comes with a rational explanation and that explanation, usually expressed in lofty terms, is only dismissed as irrational if the violence fails. Rudd said:

> The people of the Third World, including the Vietnamese and the blacks at home, were fighting to get out from under American control. How could we as white Americans just stand by and cheer them on? It would be racist to only do that. So we had to bear some of the costs by "picking up the gun."

How much this sounds like John Kennedy's Cold War pledge, from his January 1961 inaugural address—"We shall pay any price, bear any burden . . . in order to assure the survival and success of liberty"? How far is this from answering the call to free the slaves, smash the Hun, rid the world of fascism? Is this not another Urban II speech about our obligation to take up arms to drive out evil?

But another lesson from the Weather Underground is that violence is a virus that infects. Both the movements and their enemies understood that if violence could be planted in a group, it would spread and eventually destroy the movement. Richard Gregg had warned: "Every 'blood and iron' type of governor fears nonviolent resistance so much that he secretly hires so-called 'agents provocateurs' who go among the nonviolent resisters pretending to be of them, and invite them to deeds of violence. . . ." SDS and the other movements of the 1960s were laced with these undercover agents urging violence, trying to infect the movement with it. In Mexico City in 1968 the government filled trash barrels with rocks and planted them at the sites of demonstrations. In the movements in most countries, when someone proposed violence that person was immediately suspected of being an undercover agent.

But in the end, the civil rights movement, SDS , and many other nonviolent movements around the world were attacked, seduced, and destroyed by what Gandhi called the *goondah raj,* the rule of thugs. They could have learned from the thirteenth-century Cathars. Only if the nonviolent side has the discipline to avoid slipping into violence does it win. It is the tactic of state thuggery to reduce the dialogue to the level of thugs. That is why the police kept beating Bayard Rustin in 1942, to lure him into responding, thereby defeating him. The moment the nonviolent adversary accepts violence, as the Black Panthers did when they announced that they would respond to violence with violence, then it has been conceded that violence is acceptable, and it is only a question of who has the greater physical force. Hannah Arendt wrote: "The practice of violence changes the world, but the most probable change is a more violent world."

Rudd now sees the entire Weathermen adventure as "self defeating":

> We played into the hands of the FBI by destroying the above-ground organization we had come to control, SDS, and by isolating ourselves as "terrorists." We also helped create division in the anti-war movement. We might as well have been FBI agents, we did their work so well.

In 1968, before the Weathermen were even founded, Richard Nixon was elected president of the United States, beginning a backward march from civil rights and progressive ideas that has continued for almost half a century. The main thrust of his election campaign was law and order, a nakedly cynical appeal to the fear that was gripping Americans. They were afraid because of the violence that had pervaded, a violence that the state itself had played a huge part in. But the lesson remains that violence provokes fear, and people who are afraid will rarely act well.

In 1988, Rudd got another remarkably concise lesson on the subject, from the poet Allen Ginsberg. They were sharing a car ride after a radio interview about the twentieth anniversary of the Columbia University student uprising, and Rudd asked Ginsberg where he thought the Weathermen had gone wrong. Ginsberg, the Jewish Buddhist, said, "A lack of *rachmonis*"—the Yiddish word for compassion.

———

The world in general, and especially the U.S. military, learned many lessons from Vietnam. Feeling that its ability to make war was in jeopardy, the U.S. government labeled its failure a syndrome, the "Vietnam syndrome"—as though it were a disease. The "Vietnam syndrome" was, and it still exists in the military, a healthy hesitance to go to war. Most films and books about the Vietnam War neither romanticized nor glorified war, the way World War II movies did. It is clear that movies play a role in shaping attitudes about war. In Vietnam, American noncommissioned officers often complained that they lost men because they would kneel rather than staying down low in a firefight, as they were trained to do. The reason some young soldiers ignored their training was that years of World War II movies had trained them otherwise. They came to call this often fatal firing position "John Wayneing," after the famous war-movie actor who had never been in combat.

After seeing Michael Cimino's painful 1979 film *The Deer Hunter*, Jan Scruggs, a wounded and decorated Vietnam veteran, was inspired to form a veterans' organization to raise money for a Vietnam War memorial monument in Washington, D.C. The monument they erected, like most of the Vietnam War films themselves, represents a break with the tradition of celebrating warfare. The selected design, by Maya Ying Lin, was simply a wall of polished stone with the names of the more than 58,000 people who were killed in that war. The simple realities of presenting all those names in the order they were killed stated the painful reality that this is what war is, the large-scale killing of individuals.

But by the time the veterans' group had the design ready, Ronald Reagan was president. Reagan didn't like the "Vietnam syndrome" and he didn't like the monument design. Critics, including some veterans, complained that it was "flagless." Secretary of the Interior James Watt refused to issue a building permit until a traditional romantic statue of three servicemen was allowed to be erected next to it. Not everyone was ready for Vietnam's lessons.

XI

RANDOM OUTBREAKS
OF HOPE

We may make contact with ambitious species on other planets or stars; soon thereafter there will be interplanetary war. Then, and only then, will we of this earth be one.

—WILL AND ARIEL DURANT,
The Lessons of History, 1968

No doubt it seemed that way in that most violent year, 1968. The Durants calculated that year that of the previous 3,421 years, only 268 had been without war. An outbreak of world peace has not been experienced since then. But the world is not without hope. Just as most news media, political leaders, cultural institutions, and pundits ignore nonviolence and glorify war, the world does not often recognize triumphs of nonviolence.

Most of the coverage of political movements in the 1970s focused on the waning of the activism and idealism of the 1960s. In France, West Germany, Italy, and Mexico, the main movements dispersed and the small, violent Weathermen-like groups went underground. The governments and police forces of these countries spent the next two decades trying to root out and kill every one of these underground "revolutionaries."

But not everyone went home or underground. A great many used the nonviolent techniques they had learned and applied them to further the antinuclear cause, women's rights, the environmental movement, and gay rights, an issue so repressed and unpopular that even Bayard Rustin, a lifelong fearless rebel and himself a homosexual, did not embrace it as a cause until the 1980s.

Most of the Mexican opposition did not go underground but instead, for the first time in Mexico's violent history, opposed the government without violence, challenging rigged elections, exposing fraud and embezzlement among government leaders, and gradually putting pressure on the political process, forcing it to open up little by little, until in the year 2000 a fair Mexican election resulted in the peaceful removal from power of the long-entrenched ruling party.

In the winter of 1989, when democratic yearnings began to unseat the Polish Communist Party, when all of Central Europe fell, and finally the Soviet Union was dismantled, the world was shocked and completely taken by surprise by this sudden turn of events. No one was more caught by surprise than the befuddled U.S. president, Ronald Reagan, and his advisers. Had they done it? In time they decided they had—that they had overthrown the Soviet Union by taking a hard line. Of course U.S. governments had been taking a "hard line" ever since the Russian Revolution. Woodrow Wilson had even invaded. But Ronald Reagan, by being a good cold warrior and stepping up the nuclear arms race, had pressured the Soviets right out of existence. To make this claim—and some still make it—is to ignore the Eastern Europeans who dedicated their lives to slowly, nonviolently chipping away at Soviet authority. And so one of the most spectacular victories of nonviolence in history is seldom mentioned.

A cynical observer would say that Reagan and the Reaganites were engaged in a standard exercise of political deceit. It is better to take credit than admit being caught by surprise. Some call this leadership. But the Reaganites may in fact have been sincere. It may have been inconceivable to them that the Soviet Union could have been brought down by nonviolence. Like Reinhold Niebuhr, most people believe that nonviolence can never work against a ruthless dictatorship. And Reagan had already labeled the Soviet Union "the Evil Empire." Either it was not as ruthless and evil as he had claimed, or nonviolence is capable of destroying an evil empire.

———

It all began in 1965, at the University of Warsaw, with two Communist students, Jacek Kuroń and Karol Modzelewski, who, though arrested and expelled from the party, continued to organize among the Communist elite at the university. In 1968, students in Warsaw began demonstrating against the authorities after they closed down a stage production by the National Theater Company. The students were good Communists from good Communist families who believed that marching in protest was something a good Communist did. It came as a soul-altering shock when the Communist government unleashed almost unrestrained violence, beating them in the street. The young Communists, who had simply wanted to reverse a bad decision on censorship, began rethinking their entire relationship to the regime. The movement spread throughout Poland and was met everywhere with sufficient violence that, by March, it appeared to have been crushed. These students were an elite and had failed to interest the rest of the population in their issues of intellectual freedom.

Adam Michnik, who had been a leader of the student movement, set out to unite students and workers through KOR, the Polish acronym for the Committee for the Defense of Workers, which offered moral and material support to dissident workers. The Polish government was trying to revitalize the economy through a series of measures that forced workers to produce more for less pay. The government harassed, arrested, and beat dissidents, but KOR continued its work. After a dock strike in Gdansk in 1980 led to the creation of a dissident trade union movement—Solidarity—the goal had been accomplished—a unified movement of intellectuals and workers. They even brought in the Church, and all of these factions were committed to the principles of nonviolence, consciously embracing Gandhi's belief in the power of nonviolent noncooperation and never doubting that it would work for them.

But the rest of the world had no faith in nonviolence. The British in particular, avoiding self-discovery, always insisted that Gandhi had proven not the effectiveness of nonviolence but rather the essential civility of the British—or, as Stokely Carmichael liked to say, the Brutish Empire. In a 1949 essay on Gandhi in the *Partisan Review*, George Orwell even specified that Gandhi's tactics would never work against the Russians. In a 1985 essay, Czesław Miłosz wrote about how frequently people make the argument that Gandhi's approach could not work against the modern totalitarian state, to which he replied, "Our natural tendency to place the possible in the past leads us often to overlook the acts of our contemporaries, who defy the presumably unmovable order of things."

The movement's secret weapon was patience. The Poles accepted that it would take time. Jan Litynski, who had been a student protester in 1968, for which he spent two and a half years in prison, after which he was active editing underground newspapers and organizing labor, told the *New York Times*, when the end came, "I guess what surprised me most is that I just did not think it would happen so soon." It had taken twenty years.

———

Adam Michnik, writing from prison in 1985, gave several arguments in favor of nonviolence. Michnik, who grew up in the repressive society created by the Russian Revolution, wrote: "Taught by history, we suspect that by using force to storm the existing Bastilles we shall unwittingly build new ones." He said, "In our reasoning, pragmatism is inseparably intertwined with idealism." And this approach may be characteristic of all successful nonviolent activists. It is what perplexed Orwell about Gandhi. Michnik did not believe violence was a viable option for his cause. As he put it simply: "We have no guns."

James Madison said, "All governments rest on opinion," and this is no less true of dictatorships than democracies. The problem with dictatorships is that the leadership is more corrupted by power than that of the democratic tyrant who can be voted out. So while the Soviet Union worked hard at maintaining public opinion, if it felt challenged, it usually responded brutally, even though this was unpopular. By the end of the 1980s such a large part of the population had turned against it that the Soviet Union could no longer function. On October 7, 1989, East German Communist Party leader Erich Honecker ordered security forces to open fire on demonstrators in Leipzig. Egon Krenz, his man in charge of security, flew to Leipzig to prevent the shooting. Krenz feared that if their security forces opened fire it would mean the end of the regime. Ten days later, after Honecker was forced to resign, the regime did resort to violence. Within a month they were gone and the Berlin Wall was being chipped away by souvenir hunters.

In Prague, on November 17, 1989, students marching in a procession to commemorate the shooting of a student by the Nazis were attacked by the Communist police. With East Germany falling, the Czech regime believed they needed to make a show of force. Rallies protesting the regime grew in numbers every day after the attack. The regime lasted only a few more weeks.

In Czechoslovakia it had begun twenty-one years earlier with a dedicated Communist who, like many on the left, was committed to nonviolence. Alexander Dubček was tall and dull, seemingly the opposite of a Gandhi. When the forty-six-year-old Slovak party hack became leader of the Czechoslovakian Communist Party on January 5, 1968, no one expected anything out of the ordinary, which was the reason the Russians had backed him. His wife and two sons sobbed uncontrollably at his fate. But he came with a lot of ideological underpinnings. His parents were, in Dubček's words, "Slovak socialist dreamers," who had met after immigrating to Chicago's North Side. His father was a pacifist who opposed World War I and had gotten involved with Quakers who tried to smuggle him to Mexico to avoid military service. He was caught and sent to prison. After the war, disenchanted with the United States and excited by the breakup of the Hapsburg Empire, they returned to the new state of Czechoslovakia, where their son Alexander was born a few months later. No doubt as Alexander's career advanced, the Communists took into account his pedigree, that his parents were Marxists who had worked for the Soviet Union almost from the beginning and that Alexander had grown up on the Soviet frontier, trying to promote cooperative farming among the tribesmen. What apparently no one had considered was that Alexander had been raised not simply by good Marxists, but by that most dangerous breed—pacifists.

By the time Czechoslovakia was placed under his quiet control, it had become the most repressive state in the Eastern Bloc, having missed out on post-Stalinist reforms. In 1968 the Czechoslovakians were yearning not for a dull, gray leader but for dramatic change. In public meetings Dubček asked people what they wanted. They told him, and so in the midst of the totalitarian empire began a bloodless revolution. Freedom of speech, a free press, freedom to travel were all part of Dubček's "Communism with a human face." Prague became the place to visit, and the limited hotel rooms were always booked. But Dubček wanted neither to overthrow nor to revolt. He wanted merely to reform. He repeatedly made clear to Moscow that Czechoslovakia did not want to withdraw from the Warsaw

Pact and regarded the Soviet Union as its close ally. He was determined not to repeat the mistakes of Hungary in 1956. The Hungarians had not chosen nonviolent reform but armed revolution, and the Soviets went in to crush it. The invasion, though unpopular, was not firmly opposed, because the world accepts the idea of using violence against violence.

But Czechoslovakia was different. In August, when the Soviets invaded with tanks, Dubček urged his people not to resist, despite the fact that the Czech army was considered the best in the Warsaw Pact. When the world saw the Soviet Union invade one of its closest allies, and saw its tanks stared down by unarmed students, its defeat had already begun. Mikhail Gorbachev, the last Soviet leader, years later, after his country had collapsed, agreed that nothing was ever the same after the 1968 invasion.

Gorbachev was part of a delegation that visited Czechoslovakia in 1969 to try to win over Communist youth. He encountered hostility everywhere. Often party officials seemed afraid to be in contact with the Soviet delegation, not afraid of violence but afraid to lose all standing with Czechoslovakians. "People refused to talk to us and did not answer our greetings. It was very unpleasant," he recalled. In Bratislava, Slovak workers refused to meet with him. "I returned home weighed down by gloomy thoughts," Gorbachev wrote in his 1995 memoirs.

With violence or without, it was too late for the Soviets. They had lost all credibility with the people. La Boétie had been right in the seventeenth century—you just have to stop conniving with the thief who plunders you. Gandhi, making the same point, had written, "No government can exist for a single moment without the cooperation of the people, willing or forced, and if people withdraw their cooperation in every detail, the government will come to a standstill." This became the successful strategy of Poland, Hungary, and Czechoslovakia.

Perhaps the reason the world did not appreciate what was happening in these countries was that part of the strategy, especially in Czechoslovakia, was to focus on small victories of everyday life. Václav Havel, the Czech dissident playwright and later president of the Czech Republic, called it "defending the aims of life." Organizations were formed to support the families of those persecuted by the government; alternative "universities" taught the things excluded from official education; environmental groups were formed and cultural activities established. Even before Solidarity, in Poland, alternative trade unions were created. Increasingly citizens could live life apart from the one established by the regime. Though the actions were small, the goals were large. Havel called this "living within the truth" and argued that if people lived their lives parallel to the state system and not as a part of it—which he termed "living within a lie"—there would always be a tension between these two realities and they would not be able to permanently coexist. The answer to the abuses of state was not to participate.

Gorbachev, faced with such nonviolent parallel movements, and having learned the lesson of the disastrous invasion of 1968, had no choice but to compete by creating his own reforms. Not surprisingly, after years of struggle, people found his less appealing than those of the homegrown dissident movement. As Michnik put it in his paraphrase of Marx, "A nation, like a woman, cannot be forgiven for one moment of oblivion when she allows a villain to take possession of her."

———

The Soviet Union and Mexico were only two of numerous infamously ruthless regimes that were overthrown by the tactics of nonviolence toward the end of the blood-besotted twentieth century.

In 1977 a small group of women—only fourteen originally—took on one of the most ruthless and brutal dictatorships in the world, the Argentine military junta. The women, dressed in comfortable flat shoes so that they could run for their lives if they had to, had originally gathered at Buenos Aires's Plaza de Mayo, in front of the government building, the Casa Rosada. They were the mothers of just a few of the thirty thousand Argentines who "disappeared" into the blue Ford Falcons of the military government, never to be heard from again. What could be done in the face of this kind of ruthlessness—a regime that even kidnaped pregnant women, waited for the babies to be brought to term, then murdered the mothers and gave the children to friends of the regime? By the end of the year there were weekly demonstrations by 150 women who called themselves "Las Madres de Plaza de Mayo." They placed ads in newspapers and circulated petitions. When the military started making arrests, even more turned out. Other dissident organizations started to connect with Las Madres. Soon there was a network of them. In February 1978 the police beat them in the plaza. After that, they turned up every Thursday by the hundreds. Many factors led to the collapse of the Argentine regime in 1982, just as there would be many factors involved in the fall of Marcos in the Philippines, and of the Soviet Union. But these women were one of the important catalysts for change.

In the mid-1980s, when the world looked at the Philippines, it saw the corrupt regime of Ferdinand Marcos engaged in warfare with the Marxist New People's Army in the north and the Muslim separatists of Mindinao in the south. Few even noticed in 1985 when Bishop Francisco Claver declared, "We choose nonviolence merely as a strategy for the attaining of the ends of justice, casting it aside if it does not work." Since violence had failed to topple the regime in years of fighting, most observers assumed nonviolence would not go far. Eight months later, Marcos had been overthrown by "people power."

The history of the struggle to emancipate people of color in South Africa is one of shifting tactics between violence and nonviolence. Gandhi's campaign at the beginning of the twentieth century was nonviolent. The African National Congress, modeled on Gandhi's Indian National Congress, was also nonviolent and focused on fighting legal battles. After 1948, when an openly racist National Party won elections, a new generation of militants, led by Nelson Mandela, energized the old movement, while remaining committed to nonviolence. They went to jail for deliberately defying laws of segregation. But by 1953 some demonstrations turned into riots and the government passed laws approving such practices as whipping protesters.

In 1960, to protest the requirement of black people to carry official passes, thousands showed up at once in police stations to be arrested for not having them. At an industrial city in the Transvaal, Sharpeville, the police responded by opening fire, killing almost seventy. Nonviolence was getting more difficult to believe in, and Mandela formed an armed wing of the ANC and began a bombing campaign, which led to the banning of the ANC and other organizations as well as the imprisonment of their leaders. Placed on trial in 1964, Mandela said that the nonviolent policies of the ANC over the past fifty years had been an utter failure. He said, "As violence in this country was inevitable, it would be unrealistic and wrong for African leaders to continue preaching peace and nonviolence at a time when the Government met our peaceful demands with force."

Nonviolence seemed over for South Africa. Even Kenneth Kuanda, the Rhodesian-born president of Zambia and the most prominent voice for nonviolence on the continent, changed his mind. "It is impossible," Kuanda concluded, "to be nonviolent in an unjust society." Kuanda hastened to add that Africans must not consider the fight to end apartheid a "Holy War," that instead it was a "messy, brutal, degrading business." He did not wish to decorate this violence with the bogus notion of a "just war," which he likened to trying "to disinfect violence by pouring over it a balm of sweet scented theology."

Kuanda had not lost his conviction that violence was wrong, only his faith that nonviolence could work. But it was becoming increasingly obvious that violence wouldn't work either. The white government had all the guns. In pitched street battles, the blacks always sustained the bulk of the casualties. At the height of the violence a new voice emerged, the Anglican bishop Desmond Tutu. Delivering to an angry black audience the funeral oration for Steven Biko, a black leader murdered in prison, Tutu said: "Pray for the leaders of this land, for the police—especially the security police and those in prison service—that they may realize that they are human beings too. I bid you pray for whites in South Africa."

Black resistance was changing. A new generation of leaders, such as Mkhuseli Jack, believed violence alienated too many people from the movement. They thought they would never win by fighting the police in the black townships: they instead had to reach the whites "where they lived," and the way to do that was with an economic boycott. When white store owners lost a third of their business, the white merchant class began to withdraw its support of the apartheid regime. In December 1989, South African president F. N. de Klerk began negotiating with the still imprisoned Mandela. Two months later Mandela was released; four years later he was elected president of South Africa.

Throughout the 1980s, the ANC had increased its violent attacks, from 13 in 1979 to 281 in 1989, so it can still be argued that violence played a role in the overthrow of apartheid. Given such events as the Sharpesville massacre, it is impossible to say whether more people died in violence than would have in nonviolence. One campaign of several months in 1976 resulted in 1,149 deaths, all but 5 black. Even most ANC members agree that their actions were not nearly as effective as boycotts and economic sanctions, and that what finally brought de Klerk to negotiations with Mandela was the understanding of both men that they could not achieve their goals through violence.

One place in which violence has been a complete and enduring failure is the Middle East. Ami Ayalon, a small, fit-looking man who was a former commander of the Israeli navy and former director of Shin Bet, Israel's internal security service, said, "If you sat on another planet and watched the world your conclusion would be that violence is not working. That is not the conclusion people come to in the Middle East." In a 2005 interview he said, "Violence doesn't work but it is very difficult to prove it. A poll six weeks ago showed that most Palestinians, about 75 percent, believe that the intifada succeeded. They believe that we understand only the language of force. Most Israelis believe that we won because Palestinians understand only the language of force."

In the Israeli-Palestinian conflict, force is credited for everything. When the Palestinians said they were willing to recognize the right of Israel to exist, many Israelis assumed they had been forced into the concession by the presence of Jewish settlements.

When the Israelis removed Jewish settlements from Gaza, many Palestinians said that it was because of the intifada, the Palestinian campaign of confronting Israeli forces with rock-throwing youths, and the suicide bombers. Some pacifists like to point to the intifada as an example of nonviolence working. This is problematic for two reasons, the first being that rock-throwing is not nonviolence, it is simply low-technology. The other is that the Israelis over the years have not become a kinder, more understanding people but a more embittered, hard-line, and militaristic one. Being attacked has done nothing to improve them. Once more the ideals of a religion have been tossed away for the demands of a state. A good number of the leading figures in Israeli politics during the country's second twenty-five years would have been deemed completely unacceptable right-wing militarists during the country's first twenty-five years. Israeli and Palestinian leadership has become like an old marriage where both parties are starting to look like each other.

The state of Israel, the "Constantine" of Judaism, shoves aside traditional Jewish values for traditional statist values, while the Palestinian leadership perverts the meaning of Islam, the concept of *jihad,* and, most especially, the concept of martyrdom. Mohammed had detailed specific criteria for martyrdom, the true *jihad* that was a gateway to heaven. He preached that those who enriched themselves or pleased their vanity would not be recognized by God. And suicide was simply cheating, trying to fake your way into heaven.

There are Islamic specialists who have spoken out against these abuses. And people like Ami Ayalon, both Israelis and Palestinians, have become frustrated with political leadership. Ayalon believes that most Israelis and Palestinians are ready to make the necessary concessions for peace and that political leadership is the only obstacle. This is one of several Israeli-Palestinian peace movements that are much talked about but are still far from gaining prominence in the Middle East.

It is difficult to look at the Middle East and believe that the world is making much progress. It was hard to listen to the violent Palestinian group Hamas, or to the violent Islamist group Al-Qaeda, which spoke the doctrine of the thirteenth-century Ibn Taymiyah, or to George W. Bush, who spoke of a crusade and a permanent "war against terrorism," steeped in the rhetoric and logic of Urban II, and believe that the world has made much progress since the eleventh century. As the Polish poet Czesław Miłosz pointed out, "The terrorism of revolutionaries and the terrorism of the state seem to be two faces of the same coin." Both sides claim that God is on their side, but the god cited is a god of killing not found in the religions of either side. "The issue," said Kenneth Kuanda, "is not whether God is on our side, but if we are on God's side." Still, in light of modern history, it is getting ever more difficult to argue that nonviolence cannot work.

———

In October 2002, by a vote of 77 to 23 in the Senate and 296 to 133 in the House of Representatives, the U.S. Congress voted to give President George W. Bush the authority to attack Iraq because it was building "weapons of mass destruction." It is a peculiarly accepted notion that the United States, the only country ruthless enough ever to have used atomic weapons—and used them against a civilian population—should be trusted with a monopoly on weapons of mass destruction. But worse, the claim of Iraqi weapons was a blatant lie contradicted by the United Nations weapons inspector in Iraq, among many other reliable sources. But that isn't to say that the world isn't making progress. In 1964, President Lyndon Johnson went to Congress with another blatant lie—what war doesn't have its founding lies?—about an alleged attack by the North Vietnamese, and Congress gave him the authority to attack Vietnam, the House voting unanimously in favor and the Senate concurring with only two dissenting votes.

Every war produces a fresh crop of peace activists with a desire to change the world and a fresh determination to do it without violence. And for them every new war is a setback. But the advocates of peace and nonviolence come back stronger and more numerous each time. Given this formula, with enough wars the world may yet find peace. But that seems a long way off. John Stuart Mill, the nineteenth-century British philosopher, wrote in his 1859 master-

piece *On Liberty:* "But, indeed, the dictum that truth always triumphs against persecution is one of those pleasant falsehoods
which men repeat after one another till they pass into common-
places, but which all experience refutes." But Mill added, "the real
advantage which truth has" is that successive generations keep re-
discovering it, until "from favorable circumstance it escapes persecution until it has made such head as to withstand all subsequent
attempts to suppress it."

The great success of the anti–Vietnam War movement was its
ability to get the Vietnam War officially recognized as an unpopular war. The War of 1812 and World War I were probably equally
unpopular, but this was covered up once those wars ended. One result of Vietnam's status as an unpopular war was that its veterans
were not forced to play heroes. They were, in fact, largely ignored.
Psychiatrists believe that the worst thing for a combat veteran is for
him not to be allowed to talk about what is bothering him. The
hero's welcome of 1945 and the silence of 1973 both had that effect.
But because the Vietnam veteran was not celebrated, he had nothing to hide. The hero of World War II was not permitted to acknowledge guilt and shame and cold sweats as he reenacted combat
in his mind every night. The Vietnam veteran could be open about
it. And because of this the trauma of Vietnam veterans has informed the Korean War and World War II veterans and also helped
Gulf War and Iraq War veterans. What became widely publicized
as post-traumatic stress disorder, or PTSD, it was discovered, is not
unique to Vietnam veterans. A Boston psychiatrist specializing in
PTSD in Vietnam veterans, Jonathan Shay, wrote a book titled
Achilles in Vietnam in which he observes that most of the issues he
treated in his patients were mentioned by Homer in the *Iliad*. Shay
wrote of the Greek concept of *thémis*—"what's right." He noted,
"The specific content of the Homeric warriors' *thémis* was often
quite different from that of American soldiers in Vietnam, but what
has not changed in three millennia are violent rage and social withdrawal when deep assumptions of 'what's right' are violated."

The miracle is that despite films, books, television, toys, plaques
on buildings and monuments in parks, lessons in school, and the

encouragement of parents, most human beings enter the military with a strong disposition against killing. It turns out that we are not built for killing other people. The purpose of basic training is to brainwash out this inhibition and turn the recruit into an efficient killer.

Military training teaches survival, it teaches loyalty to the group—"your buddies." Sometimes training techniques become bizarre. Ken Hruby, a retired colonel who served in both Korea and Vietnam, said that when he was a West Point cadet, he and his classmates were given ballroom dancing instruction with attractive young women brought up from the Arthur Murray Dance Studio in New York City. Often after the dance class the cadets would be ordered into fatigues and sent out for bayonet drill. He observed that dance and bayoneting used "the same circular choreographic notation," that the steps to a waltz and "the vertical butt stroke series" were remarkably similar.

The training usually works well enough so that the soldier does what he has to do to survive, though he not infrequently, if unobserved, will fail to kill.

In 1947 Samuel Marshall, popularly known as "Slam," upset the entire U.S. military establishment with a small book titled *Men Against Fire: The Problem of Battle Command on Future War.* Marshall served as the chief U.S. Army combat historian during both World War II and the Korean War. In his 1947 book, he claimed that at best one in four U.S. combat soldiers in World War II ever fired their weapon and in most combat units only 15 percent of the available fire power was used. Basic training had failed to turn most recruits into killers.

In 1946, after the end of World War II, calling for an end to war, Albert Einstein said, "A new type of thinking is essential if mankind is to survive and move to higher levels." Victor Hugo's dream of a United States of Europe is now unfolding and Europe for the first time in history is the most antiwar of continents. But it still remains to be seen if the new Europe will be a force for peace, the new kind of entity that Hugo wrote of, or merely the emergence of another superpower playing the power politics that lead to war. The appearance of a European flag and a European national anthem, and discussions on European armed forces, all the trappings of the old warring nation-state, are among the more troubling signs.

There have been few wars in history that have had as many opponents as the American invasion and occupation of Iraq. That opposition rose up far more quickly than the opposition to the Vietnam War had. And it is getting more difficult to find people to fight wars. Most of those who join the military do so in search of economic opportunities and hope that they never have to go to war. Every time a voice is heard of the privileged, a politician, a business leader, Major League Baseball, Hollywood, one of the millionaires who present television news, saluting "the courage of our fighting men and women," who they claim—often against all logic—are "defending our freedom," listen closely and you will hear a familiar strain from that old-time tune pointed out by Petr Chelčický in the fifteenth-century Czech lands: the rich bamboozling the poor.

One of the few lessons the U.S. government seems to have retained from the Vietnam experience is to avoid a military draft and keep college campuses quiet. The U.S. all-volunteer army is nothing more than a draft of the poor, the disadvantaged, and the unemployed. Recruiters are unscrupulous in avoiding the subject of war, instead promising education and job training. In November 2002, Clifford Cornell of Arkansas, three years out of high school and jobless, walked into an army recruitment office in a shopping mall. Interviewed in Toronto in 2005, he said, "I didn't know anything about Afghanistan or the possibility of going to Iraq. I guess that's what I get for not following the news."

The army recruiter had promised that he would never be sent overseas. "His job is to lie," said Cornell. "I have learned through two years in the military that most of what they tell you is a lie. They have to lie to get people to sign up. . . . All I heard is 'If you sign up for this you get a $9,000 bonus.' "

In the summer of 2004, when Cornell learned he was being shipped out to Iraq, he went to his sergeant and said, "I'm not supposed to be shipped overseas. The recruiter said so." The sergeant just laughed. To Cornell the problem was clear. As he put it, "One:

I don't believe in war. Two: I don't agree with thousands of innocent people being killed." He drove to Toronto and deserted. "It was not an easy choice," he said. "You are trained in the military to have loyalty to your buddies."

Jeremy Hinzman, a volunteer from Rapid City, South Dakota, enjoyed the military. "I loved every part of it," he said. "Except the big picture. I mean, shooting a $5,000 rocket that can only be fired once—it doesn't get any better than that." When he applied for conscientious-objector status, he was asked if he would defend his family against a violent intruder. He said that he would, and his application was denied.

"I would like to get to a state where I could say no to this, too," said Hinzman. "But isn't saying no to war of value in itself? Do you have to be a perfect human being to reject war?" After a tour of duty in Afghanistan and shortly before being sent to Iraq, Hinzman deserted and moved with his Vietnamese-American wife to Toronto. He said, "If you are shot in the head you don't know the difference, but if you kill a bunch of people, you have to live with it. When you put someone in your sights and pull the trigger, you cross a line." He wondered why there was not wholesale desertion. "Am I in a paradigm that is only mine?"

The early-twentieth-century French novelist Anatole France wrote: "War will disappear only when men shall take no part whatever in violence and shall be ready to suffer every persecution that their abstention will bring them. It is the only way to abolish war." And as William Penn said in the seventeenth century, "Somebody must begin it."

One of the greatest lessons of history is that somebody already has.

THE TWENTY-FIVE LESSONS

1. There is no proactive word for nonviolence.
2. Nations that build military forces as deterrents will eventually use them.
3. Practitioners of nonviolence are seen as enemies of the state.
4. Once a state takes over a religion, the religion loses its nonviolent teachings.
5. A rebel can be defanged and co-opted by making him a saint after he is dead.
6. Somewhere behind every war there are always a few founding lies.
7. A propaganda machine promoting hatred always has a war waiting in the wings.
8. People who go to war start to resemble their enemy.
9. A conflict between a violent and a nonviolent force is a moral argument. If the violent side can provoke the nonviolent side into violence, the violent side has won.
10. The problem lies not in the nature of man but in the nature of power.
11. The longer a war lasts, the less popular it becomes.
12. The state imagines it is impotent without a military because it cannot conceive of power without force.
13. It is often not the largest but the best organized and most articulate group that prevails.

14. All debate momentarily ends with an "enforced silence" once the first shots are fired.
15. A shooting war is not necessary to overthrow an established power but is used to consolidate the revolution itself.
16. Violence does not resolve. It always leads to more violence.
17. Warfare produces peace activists. A group of veterans is a likely place to find peace activists.
18. People motivated by fear do not act well.
19. While it is perfectly feasible to convince a people faced with brutal repression to rise up in a suicidal attack on their oppressor, it is almost impossible to convince them to meet deadly violence with nonviolent resistance.
20. Wars do not have to be sold to the general public if they can be carried out by an all-volunteer professional military.
21. Once you start the business of killing, you just get "deeper and deeper," without limits.
22. Violence always comes with a supposedly rational explanation—which is only dismissed as irrational if the violence fails.
23. Violence is a virus that infects and takes over.
24. The miracle is that despite all of society's promotion of warfare, most soldiers find warfare to be a wrenching departure from their own moral values.
25. The hard work of beginning a movement to end war has already been done.

ACKNOWLEDGMENTS

I have to thank those who said no, and those whose brave examples taught me about nonviolence when I was young and most needed the guidance; most particularly Dave Dellinger and the Mobilization Committee to End the War in Vietnam, the "Mobe," which first gave me and so many other people a voice. Also thanks to Mark Rudd; to Matt Meyer; to Ralph Digia, who has been an example to people for almost a century; to Alastair Upton, for his help in the archives of Charleston in Sussex, England; and to Tom Hayden, who took the time to disagree with me.

Thanks to my editor, Nancy Miller, for her careful, thoughtful work, and my agent, Charlotte Sheedy, for her constant support. Thanks to Susan Birnbaum for a thousand things I could not have gotten done without her help. Most especially thanks to Marian for standing, even on occasion marching, with me, and for helping to teach our child what a bumper sticker should say: NO FIGHTING!

BIBLIOGRAPHY

Ackerman, Peter, and Jack DuVall. *A Force More Powerful: A Century of Non-violent Conflict.* New York: St. Martin's Press, 2000.

Adams, David. *The American Peace Movements: History, Root Causes, and Future.* New Haven, Conn.: The Advocate Press, 1986.

Albert, Michael, and David Dellinger, eds. *Beyond Survival: New Directions for the Disarmament Movement.* Boston: South End Press, 1983.

Arendt, Hannah. *On Violence.* New York: Harcourt, Brace & World, 1969.

Armstrong, Karen. *Islam: A Short History.* New York: Random House, 2000.

Atkin, Jonathan. *A War of Individuals: Bloomsbury Attitudes to the Great War.* Manchester: Manchester University Press, 2002.

Beechey, James. "Clive Bell: Pacifism and Politics," in *The Charleston Magazine,* issue 14, autumn/winter 1996, 5–13.

Bell, Clive. *Peace at Once.* Manchester and London: The National Labour Press, 1915.

Bender, Wilbur J. *Nonresistance in Colonial Pennsylvania.* Scottdale, Penn.: Mennonite Press, 1934.

Berman, Paul. *Terror and Liberalism.* New York: W. W. Norton & Company, 2003.

Bonney, Richard. *Jihād: From Qur'ān to bin Lāden.* New York: Palgrave Macmillan, 2004.

Bownas, Samuel A. *An Account of the Life, Travels, and Christian Experiences in the Work of the Ministry of Samuel Bownas.* London: Luke Hinde, 1756.

Branch, Taylor. *Parting the Waters: America in the King Years, 1954–63.* New York: Simon & Schuster, 1988.

Brock, Peter. *Pacifism in Europe to 1914.* Princeton, N.J.: Princeton University Press, 1972.

————. *Pacifism in the United States from the Colonial Era to the First World War.* Princeton, N.J.: Princeton University Press, 1968.

————. *Radical Pacifists in Antebellum America.* Princeton, N.J.: Princeton University Press, 1968.

————. *Varieties of Pacifism: A Survey from Antiquity to the Outset of the Twentieth Century.* Toronto: University of Toronto Press, 1998.

Brokaw, Tom. *The Greatest Generation.* New York: Random House, 1998.

Bundy, McGeorge. *Danger and Survival: Choices About the Bomb in the First Fifty Years.* New York: Random House, 1988.

Cain, William E., ed. *William Lloyd Garrison and the Fight Against Slavery: Selections from The Liberator.* Boston: Bedford Books of St. Martin's Press, 1995.

Carmichael, Stokely, with Ekwueme Michael Thelwell. *Ready for Revolution: The Life and Struggles of Stokely Carmichael (Kwame Ture).* New York: Scribner, 2003.

Catchpool, Corder. *On Two Fronts: Letters of a Conscientious Objector.* London: George Allen and Unwin, 1940.

Chadha, Yogesh. *Gandhi: A Life.* New York: John Wiley & Sons, 1997.

Child, Lydia Maria. *An Appeal in Favor of That Class of Americans Called Africans.* Amherst: University of Massachusetts Press, 1996.

————. *Letters from New-York.* Bruce Mills, ed. Athens: The University of Georgia Press, 1998.

Clausewitz, Carl von. *On War.* Michael Howard and Peter Paret, eds. Princeton, N.J.: Princeton University Press, 1976.

Cooper, Sandi E. *Patriotic Pacifism: Waging War in Europe, 1815–1914.* New York: Oxford University Press, 1991.

Dawidowicz, Lucy S. *The War Against the Jews 1933–1945.* New York: Holt, Rinehart, and Winston, 1975.

DeGroot, Gerard. *The Bomb: A Life.* London: Jonathan Cape, 2004.

D'Emilio, John. *Lost Prophet: The Life and Times of Bayard Rustin.* New York: Free Press, 2003.

Dellinger, David. *From Yale to Jail: The Life Story of a Moral Dissenter.* New York: Pantheon Books, 1993.

————. *More Power Than We Know: The People's Movement Toward Democracy.* Garden City, N.Y.: Anchor Press/Doubleday, 1975.

Deming, Barbara. *Revolution and Equilibrium.* New York: Grossman Publishers, 1971.

D'Este, Carlo. *Eisenhower: A Soldier's Life.* New York: Henry Holt and Company, 2002.

Dodge, David Low. *War Inconsistent with the Religion of Jesus Christ.* Boston: Ginn & Company, 1905.

Drake, H. A. *Constantine and the Bishops: The Politics of Intolerance.* Baltimore: Johns Hopkins University Press, 2000.

Durant, Will and Ariel. *The Lessons of History.* New York: Simon & Schuster, 1968.

Dyer, Gwynne. *War: The New Edition.* Toronto: Random House Canada, 2004.

Dymond, Jonathan. *An Inquiry into the Accordancy of War with the Principles of Christianity; and an Examination of the Philosophical Reasoning by Which It Is Defended.* Philadelphia: I. Ashmead & Co., 1834.

Einstein, Albert. *The Fight Against War.* Alfred Lief, ed. New York: John Day Company, 1933.

Enz, Jacob J. *The Christian and Warfare: The Roots of Pacifism in the Old Testament.* Scottdale, Penn.: Herald Press, 1972.

Fanon, Frantz. *The Wretched of the Earth.* New York: Grove Press, 1963.

Fernea, Elizabeth Warnock, and Mary Evelyn Hocking, eds. *The Struggle for Peace: Israelis and Palestinians.* Austin: University of Texas Press, 1992.

Fischer, Louis. *The Life of Mahatma Gandhi.* New York: Harper & Row, 1983.

Gandhi, M. K. *An Autobiography, or the Story of My Experiments with Truth.* Mahadev Desai, trans. London: Penguin Books, 1983.

———. *The Essential Writings of Mahatma Gandhi.* Raghavan Iyer, ed. New Delhi: Oxford University Press, 1991.

———. *For Pacifists.* Ahmedabad: Navajivan Publishing House, 1949.

———. *Gandhi on Non-Violence: A Selection from the Writings of Mahatma Gandhi.* Thomas Merton, ed. New York: New Directions, 1965.

———. *Non-Violent Resistance (Satyagraha).* Bharatan Kumarappa, ed. Mineola, N.Y.: Dover Publications, 2001.

———. *The Penguin Gandhi Reader.* Rudrangshu Mukherjee, ed. New York: Penguin Books, 1996.

Gara, Larry and Lenna Mae, eds. *A Few Small Candles: War Resisters of World War II Tell Their Stories.* Kent, Ohio: Kent State University Press, 1999.

Garrison, William Lloyd. *The Letters of William Lloyd Garrison, Volume I: I Will Be Heard! 1822–1835.* Walter M. Merrill, ed. Cambridge, Mass.: Belknap Press of Harvard University Press, 1971.

————. *The Letters of William Lloyd Garrison, Volume II: A House Divided Against Itself, 1836–1840.* Louis Ruchames, ed. Cambridge, Mass.: Belknap Press of Harvard University Press, 1971.

————. *The Letters of William Lloyd Garrison, Volume III: No Union with Slaveholders, 1841–1849.* Walter M. Merrill, ed. Cambridge, Mass.: Belknap Press of Harvard University Press, 1973.

————. *The Letters of William Lloyd Garrison, Volume IV: From Disunionism to the Brink of War, 1850–1860.* Louis Ruchames, ed. Cambridge, Mass.: Belknap Press of Harvard University Press, 1975.

————. *The Letters of William Lloyd Garrison, Volume V: Let the Oppressed Go Free, 1861–1867.* Walter M. Merrill, ed. Cambridge, Mass.: Belknap Press of Harvard University Press, 1979.

————. *The Letters of William Lloyd Garrison, Volume VI: To Rouse the Slumbering Land, 1868–1879.* Walter M. Merrill and Louis Ruchames, eds. Cambridge, Mass.: Belknap Press of Harvard University Press, 1981.

————. *On Non-Resistance.* New York: Nation Press Printing Co., 1924.

————. *The Words of Garrison: A Centennial Selection (1805–1905) of Characteristic Sentiments from the Writings of William Lloyd Garrison.* Boston and New York: Houghton, Mifflin and Company, 1905.

Garrison, William Lloyd, and Isaac Knapp. *The Liberator,* vol. 1, no. 1. Boston: The Old South Association, 1831.

Garrow, David J. *Bearing the Cross: Martin Luther King Jr. and the Southern Christian Leadership Conference.* New York: William Morrow and Company, 1986.

Gorbachev, Mikhail. *Memoirs.* George Peronansky and Tatjana Varsavsky, trans. New York: Doubleday, 1995.

Grayling, A. C. *Among the Dead Cities: The History and Moral Legacy of the WWII Bombing of Civilians in Germany and Japan.* New York: Walker & Company, 2006.

Gregg, Richard B. *The Power of Non-Violence.* Philadelphia: J. B. Lippincott Company, 1934.

Havel, Václav. *Living in Truth.* Jan Vladislav, ed. London: Faber and Faber, 1989.

Hentoff, Nat. *Peace Agitator: The Story of A. J. Muste.* New York: Macmillan Company, 1963.

———. *The Essays of A. J. Muste.* New York: Bobbs-Merrill Company, 1967.

Hillenbrand, Carole. *The Crusades: Islamic Perspectives.* New York: Routledge, 2000.

Holmes, Robert L., ed. *Nonviolence in Theory and Practice.* Belmont: Wadsworth Publishing Company, 1990.

Hugo, Victor. *L'Avenir.* Paris: le Cadratin, 1995.

Hurwitz, Deena, and Craig Simpson, eds. *Against the Tide: Pacifist Resistance in the Second World War, An Oral History.* New York: War Resisters League, 1983.

Huxley, Aldous, ed. *An Encyclopaedia of Pacifism.* New York: Harper & Brothers Publishers, 1937.

———. "Letter from Aldous Huxley to Anthony Brooke," in *Our Generation Against Nuclear War,* vol. 2, no. 4, 74–75. Montreal: Our Generation Against Nuclear War, 1963.

Ingram, Catherine. *In the Footsteps of Gandhi: Conversations with Spiritual Social Activists.* Berkeley, Calif.: Parallax Press, 1990.

Joas, Hans. *War and Modernity.* Rodney Livingstone, trans. Cambridge, UK: Polity Press, 2003.

Judt, Tony. *The Burden of Responsibility: Blum, Camus, Aron, and the French Twentieth Century.* Chicago: University of Chicago Press, 1998.

Kaunda, Kenneth David. *Kaunda on Violence.* Colin M. Morris, ed. London: William Collins Sons & Co. Ltd., 1980.

Keane, John. *Violence and Democracy.* Cambridge: Cambridge University Press, 2004.

Kellogg, Walter Guest. *The Conscientious Objector.* New York: Boni and Liveright, 1919.

Khan, Maulana Wahiduddin. *The True Jihad: The Concepts of Peace, Tolerance and Non-Violence in Islam.* New Delhi: Goodword Books, 2002.

King, Martin Luther, Jr. *Strength to Love.* New York: Pocket Books, 1964.

King, Michael. *Moriori: A People Rediscovered.* Auckland: Penguin Books, 2000.

Knightley, Phillip. *The First Casualty: From the Crimea to Vietnam: The War Correspondent as Hero, Propagandist, and Myth Maker.* New York: Harcourt, Brace, Jovanovich, 1975.

Konrád, George (György). *Antipolitics.* New York: Henry Holt and Company, 1987.

La Boétie, Étienne de. *Anti-dictator, the discours sur la servitude volontaire of Étienne de La Boétie.* Harry Kurz, trans. New York: Columbia University Press, 1942.

Lewey, Guenter. *Peace and Revolution: The Moral Crisis of American Pacifism.* Grand Rapids, Mich.: William B. Eerdmans Publishing Company, 1988.

Lieberman, E. James. "Non-Violence vs. Pacifism: A Psychiatrist's View," in *Our Generation Against Nuclear War,* vol. 2, no. 4, 59–65. Montreal: Our Generation Against Nuclear War, 1963.

McCarthy, Colman. *All of One Peace: Essays on Nonviolence.* New Brunswick, N.J.: Rutgers University Press, 2001.

McCullough, David. *John Adams.* New York: Simon & Schuster, 2001.

MacMaster, Richard K. *Christian Obedience in Revolutionary Times: The Peace Churches and the American Revolution.* Akron, Ohio: Mennonite Central Peace Section (U.S.), 1976.

Mastnak, Tomaž. *Crusading Peace: Christendom, the Muslim World, and Western Political Order.* Berkeley and Los Angeles: University of California Press, 2002.

May, Rollo. *Power and Innocence: A Search for the Sources of Violence.* New York: W. W. Norton & Company, 1972.

Michnik, Adam. *Letters from Prison and Other Essays.* Maya Latynski, trans. Berkeley and Los Angeles: University of California Press, 1985.

Moorehead, Caroline. *Troublesome People: The Warriors of Pacifism.* Bethesda, Md.: Adler & Adler, 1987.

Motion, Andrew, ed. *First World War Poems.* London: Faber and Faber, 2003.

Muse, Benjamin. *The American Negro Revolution: From Nonviolence to Black Power.* New York: Citadel Press, 1970.

Mussey, Barrows, ed. *Yankee Life by Those Who Lived It.* New York: Alfred A. Knopf, 1947.

Muste, A. J. *How to Deal with a Dictator.* New York: Fellowship Publications, 1954.

———. "The Peace Movement 1963," in *Our Generation Against Nuclear War,* vol. 2, no. 4, 3–8. Montreal: Our Generation Against Nuclear War, 1963.

Neufeld, Michael J., and Michael Berenbaum, eds. *The Bombing of*

Auschwitz: Should the Allies Have Attempted It? New York: St. Martin's Press, 2000.

Nunn, Maxine Kaufman, ed. *Creative Resistance: Anecdotes of Nonviolent Action by Israel-Based Groups.* Jerusalem: Alternative Information Center, 1993.

Orwell, George. "Reflections on Gandhi," *Partisan Review,* January 1949.

O'Shea, Stephen. *The Perfect Heresy: The Revolutionary Life and Death of the Medieval Cathars.* New York: Walker & Company, 2000.

Pauling, Linus. *No More War!* New York: Dodd, Mead & Company, 1958.

Rae, John. *Conscience and Politics: The British Government and the Conscientious Objector to Military Service, 1916–1919.* London: Oxford University Press, 1970.

Raphael, Ray. *A People's History of the American Revolution: How Common People Shaped the Fight for Independence.* New York: New Press, 2001.

Remarque, Erich Maria. *All Quiet on the Western Front.* A. W. Wheen, trans. Boston: Little, Brown and Company, 1929.

Reynolds, David S. *John Brown, Abolitionist.* New York: Alfred A. Knopf, 2005.

Sarma, D. S. *A Primer of Hinduism.* Mylapore: Sri Ramakrishna Math Printing Press, 1981.

Schell, Jonathan. *The Unconquerable World: Power, Nonviolence, and the Will of the People.* New York: Metropolitan Books, 2003.

Scott, Dick. *Ask That Mountain: The Story of Parihaka.* Auckland: Reed Books, 2004.

Shamroukh, Ziyad Abbas, Ingrid Gassner-Jaradat, and Maxine Nunn. *Palestine and the Other Israel: Alternative Directory of Progressive Groups and Institutions in Israel and the Occupied Territories.* Jerusalem: Alternative Information Center, 1993.

Shay, Jonathan. *Achilles in Vietnam: Combat Trauma and the Undoing of Character.* New York: Atheneum, 1994.

Strong, George Templeton. *The Diary of George Templeton Strong,* Volume II: *The Turbulent Fifties,* and Volume III: *The Civil War.* New York: Farrar, Straus and Giroux, 1974.

Sutherland, Bill, and Matt Meyer. *Guns and Gandhi in Africa: Pan African Insights on Nonviolence, Armed Struggle and Liberation in Africa.* Trenton: Africa World Press, 2000.

Suttner, Bertha von. *Lay Down Your Arms: The Autobiography of Martha von*

Tilling. T. Holmes, trans. New York: Longmans, Green and Co., 1908.

Taylor, Telford. *The Anatomy of the Nuremberg Trials.* New York: Alfred A. Knopf, 1992.

Thoreau, Henry David. *Civil Disobedience.* Per Bregne, ed. Los Angeles: Green Integer, 2002.

Tolstoy, Leo. *The Law of Love and the Law of Violence.* Mary Koutouzow Tolstoy, trans. New York: Rudolph Field, 1948.

Tomkinson, Leonard. *Studies in the Theory and Practice of Peace and War in Chinese History and Literature.* Shanghai: Christian Literature Society, 1940.

Trotsky, Leon. *The History of the Russian Revolution.* Max Eastman, trans. Ann Arbor: University of Michigan Press, 1974.

Tuck, Richard. *The Rights of War and Peace: Political Thought and the International Order from Grotius to Kant.* New York: Oxford University Press 2001.

Twain, Mark. *The War Prayer.* New York: HarperCollins Publishers, 1970.

Walzer, Michael. *Arguing About War.* New Haven, Conn.: Yale University Press, 2004.

————. *Just and Unjust Wars: A Moral Argument with Historical Illustrations.* New York: Basic Books, 2000.

Washington, James Melvin, ed. *A Testament of Hope: The Essential Writings and Speeches of Martin Luther King Jr.* San Francisco: HarperSanFrancisco, 1991.

Waskow, Arthur. *Seasons of Our Joy: A Modern Guide to the Jewish Holidays.* Boston: Beacon Press, 1982.

Wilson, A. N. *Tolstoy.* New York: W. W. Norton & Company, 2001.

Wittner, Lawrence S. *Rebels Against War: The American Peace Movement, 1941–1960.* New York: Columbia University Press, 1969.

Worcester, Noah. *A Solemn Review of the Custom of War.* Boston: S. G. Simpkins, 1833.

Zinn, Howard, and Anthony Arnove. *Voices of a People's History of the United States.* New York: Seven Stories Press, 2004.

INDEX